# Simple pest and disease control

Series editor:
**John Patrick**

**Colin Campbell**

A LOTHIAN BOOK

# Foreword

The control of pests and diseases is a significant part of most gardeners' activities; however in recent years our attitudes have changed. Where control with chemicals was the accepted and normal practice, it is now treated as the last option.

In preference to chemicals we are looking at more rational means of pest control: the role of natural predators, the cultivation of healthy and vigorous plants suited to their environment and the use of our own wits as a means of identifying solutions to pest problems, for example through the use of pheromone traps.

The great advantage of using fewer chemicals is that those that are applied provide a last-resort solution which is more effective, since pests are no longer building up immunity to commonly used chemicals as they did in the past.

Colin Campbell has been at the forefront of these changes in attitude among amateur horticulturists in Australia. Well known from his work on television and radio, Colin is recognised for his delightfully humorous yet well-informed and sound presentation of information. In *Simple Pest and Disease Control* Colin, in his usual form, addresses your pest and disease problems clearly and with authority.

John Patrick

A Lothian Book
Thomas C. Lothian Pty Ltd
11 Munro Street, Port Melbourne,
Victoria 3207


First published 1994

National Library of Australia
Cataloguing-in-Publication data:

Campbell, Colin, 1933– .
Simple pests and disease control.
Bibliography.
Includes index.
ISBN 0 85091 652 6.
1. Garden pests – Australia. 2. Garden pests –
Control – Australia. 3. Plant diseases – Australia. I. Title. (Series: Lothian Australian garden series).
635.0490994

Cover design by David Constable
Illustrations by Julia McLeish
Typeset in Cheltenham and Rockwell
by Bookset Pty Ltd
Printed in Australia by Impact Printing

# Contents


Foreword by John Patrick 2

Introduction 4

Insect pests 7

The life cycle; Plant damage; The chewing insects; The sucking insects; The borers; Root bulb and stem attackers; The tissue feeders; The fruit eaters; Gall-forming insects

Control 12

Stomach poisons; Contact poisons; General-purpose compounds; Systemic poisons; Types of pesticides

Common pests and what to do about them 17

Plant diseases 35

Physiogenic plant diseases; Diseases caused by living organisms

Some common diseases and their control 43

Diseases of foliage, stems, fruit and flowers; Bacterial diseases; Lawn diseases; Soil-borne diseases; Disease prevention

Using chemicals wisely 50

Poisons; How to protect yourself; Withholding periods; Using the right quantity; Storage of chemicals; Buying agricultural packs; Resistance factor; Disposal of chemicals

Some alternatives 56

Crop rotation; Companion planting; Non-chemical treatments; Home-made recipes; Hygiene; Conclusion

Table of pesticides and fungicides 61

Further reading 63

Index 64

# Introduction

**H**aving conducted a talkback radio gardening programme for more than twelve years, I am totally convinced that there is a widely held view amongst gardeners that if something moves in the garden it must be bad and, therefore, needs to die. Similar misconceptions are held about fungi.

Much of this 'kill at all costs' mentality has been promulgated by the home garden chemical companies (of which I confess to being a part) in their endeavours to sell their pest and disease control products. The competitive nature of the business has led to some pretty hard hitting advertising campaigns that have convinced a large number of gardeners that all insects and fungi are bad and, therefore, need to be eliminated.

In this book I hope to dispel that belief and to provide a reference that will enable gardeners to make a rational decision about whether or not a particular insect or fungus has to be sprayed and, if so, to choose the most effective product for the job.

Whenever I talk at a garden club or some other gathering of gardeners about pest and disease control, I invariably ask if anyone in the audience knows how many kinds of insects there are in the world — or even in their own backyards. The answers that I get are most interesting, ranging from hundreds, through thousands, with some even suggesting that there are millions of insects. And if the feeling with which this comment is made is anything to go by, I can only assume that they are all in that person's garden!

Of course, the answer that I am looking for is simply *two*.

There are only two kinds of insects in the world — namely, good and bad — and we, as gardeners, must never forget that.

The unfortunate idea that everything that moves or flies in the garden is bad and therefore needs to be eliminated is both erroneous and harmful. There are in fact more good or beneficial insects in the garden than there are bad or harmful ones. Every time we gardeners spray an insecticide around, we invariably kill both the good and the bad.

Predators are the key to a relatively pest-free garden which does not need continual spraying. To sustain a wide range of predators, you will need to provide them with food, which means that you will have to put up with a few insect pests in your garden at some stages during the year.

The role of predators must be more widely recognised, as these are the key to a relatively pest-free garden which does not need continual spraying. In order to sustain a wide range of predators though, it is necessary to provide them with food. This means that you will have to put up with a few insect pests in your garden at some stage during the year and disregard the damage they do. If you spray the insect pests as soon as they arrive, you will also eliminate the predators' food source and, as a consequence, they will go to someone else's garden for their meals.

Once the balance between predators and pests has built up to the level where they are self-sustaining, the likelihood of damage is generally significantly reduced.

This balance between predators and pests is not going to be achieved overnight, in fact it can take quite a few years. There is an organic garden in Brisbane, owned by the Charteris family, where no chemical of any kind has been used for years and the plants growing there are as healthy as one would expect to see anywhere.

I'm quite sure this garden didn't always look as healthy as it does now, because of the time it takes to achieve that delicate balance. The fact that theirs is the only real garden in the neighbourhood might also have helped. Had there been a lot of ardent gardeners nearby spraying chemicals around indiscriminately, it would have taken the predators longer to build in sufficient numbers to effectively control the pests. The Charteris' garden is now an example of a garden working in complete harmony with nature.

A plant disease can be described as anything that interferes with the normal growth or development of a plant. The majority of plant diseases are caused by fungi, but that certainly doesn't imply that all fungi are harmful. Just as with insects, there are both good and bad fungi, and whenever we spray a fungicide we invariably kill the good with the bad. From time to time mushrooms or some other fungal growth may appear in our gardens. Whenever they do, you can bet your last dollar that there will be gardeners out there spraying fungicides madly in the mistaken belief that, because some fungi kill plants, then all fungi are bad. How sad that is, because there are more beneficial fungi than there are damaging ones.

One or other of the various beneficial fungal organisms that exist in the soil breaks down organic matter. Another of the beneficial fungi inhabits the soil close to plant roots. Pine trees have an inherent ability to extract from the soil nutrients that would be considered unavailable to other plants. The credit for this is attributed to a symbiotic association between certain fungi and the roots of associated plants, an association known as 'mycorrhiza'. A similar situation is also said to exist in regard to a number of Australian native plants, such as the *Xanthorrhoea arborea* or, as it is commonly known, the Black Boy. This mycorrhizal activity is the reason why these plants are difficult to transplant and explains why it is necessary to take as much of the native soil as possible when lifting them. By taking this native soil, you are ensuring that there is a sufficient quantity of beneficial fungi present to ensure that the mycorrhizal activity takes place.

I am one of those old-timers who was brought up on a farm, and at a time when almost everyone sprayed chemicals around as if there was going to be no tomorrow. Very few, if any, safety precautions were taken. Now, of course, I realise that these were irresponsible actions and I am totally convinced that the indiscriminate application of chemicals carried out by farmers and gardeners in those days was delinquent and, at times, dangerous. The excessive use of many of those chemicals has led us into a situation in which some less harmful chemicals are nowhere near as effective as they once were. This has led in turn to a build-up of certain other chemicals in the environment. It is absolutely essential that we adopt a responsible attitude towards the use of chemicals, man-made or organic, in gardens and on farms in the years to come.

Some of my readers may have gained the impression that I am opposed to the use of chemical pest and disease control methods in the garden, but this is not true. In recent times I have adopted what I believe to be a pretty pragmatic approach to the use of chemicals in my own garden. For example, I do not spray with insecticide unless it is absolutely necessary, and then I will always use the least toxic product available for that purpose. Where possible, I will rely on the natural predators to control the pests for me, but when the 'crunch' comes and the predators are fighting a losing battle with the pests, I will bring out the insecticide spray. I might as well have something rather than nothing.

For gardeners who don't want to spray, melons and pumpkins may be protected from the fruit fly and other fruit-eating insects by covering with small pieces of rag.

Fungi are very simple plants without chlorophyll but there are some quite beautiful specimens. This was photographed at Wairakei, New Zealand.

*Xanthorrhea arborea*, or Black Boy, is an example of a plant that has beneficial soil fungi in combination with its root system.

There is another factor that enables me to keep spraying to a minimum and that is plant nutrition. I firmly believe that a well-nourished plant is better able to withstand the ravages of pests and diseases than is a poorly fed plant. This is of course no different to us humans — if our diet is lacking we are more likely to succumb to illness. A well-nourished lawn, for example, can play host to a goodly number of lawn grubs before serious damage is apparent. Similarly a plant that is making rapid growth may be making so many leaves, that the feeding efforts of a few caterpillars go unnoticed. One of the diseases that can cause problems in papaws is called 'crinkle top'. More often than not, a well-fertilised papaw will grow through the disease and the fruit that follows will be as good as gold.

So, by maintaining an adequate fertiliser programme, it is possible at least to minimise the damage done by pests and diseases to the garden and to reduce the need for other control methods. It makes a lot more sense to spend money on plant nutrition than on pest and disease controls.

I like to think that I am acting in a responsible manner when it comes to my use of chemicals, bearing in mind that it is a lot easier to have a totally chemical-free garden in the cooler regions of Australia and New Zealand than it is in the tropics and subtropics. The pressure from pests and diseases is just so much lower in the temperate and cool regions that predators are able to keep them under control. I am, however, totally convinced that in an ideal world the use of chemicals would be unnecessary, but as yet that ideal world is a long way off. Rather than adopt a head-in-the-sand attitude towards the use of chemicals in the garden, it is my fervent wish that those gardeners who continue to use chemicals will do so with the utmost caution and responsibility.

So, if you, like me, are prepared to use chemicals, man-made or natural, let's make certain we know just what it is that we are spraying and can choose the most appropriate control measure for the task. To do this we will need to be able to identify the pests and fungi that attack our gardens and know exactly what steps we need to take to eradicate them or live with them.

In later chapters we will look at the organic alternatives to chemicals and the precautions necessary to protect ourselves.

# Insect pests

Insects are one of the most successful groups in the animal kingdom and it is estimated that there are more than one million species. At least ten thousand of these are known to be damaging to plants. Fortunately not all of them live in your area!

All plants have their own range of pests, some of which are quite specific to those particular species. This is one of the things that causes me to smile a little when I hear proponents of the Australian native plant movement expounding the easy-care virtues of their favourite plants. Now I'm not in the business of denigrating our Australian native plants, nor their devotees, but I do know that some native plants can be just as hard to look after as exotics. And it does stand to reason that if a plant has lived in this country for as long as some of our native plants have, there is going to be a range of pests that will have found it long before this.

Experience has shown that many gardeners have great trouble identifying the pest that is causing problems in the garden. Effective identification is the very cornerstone of responsible insect control; once the insect has been correctly identified, it is then possible to choose the most effective control measure. This may simply be to leave the insect alone, in the knowledge that it will either be controlled by a predator or it will simply go away of its own accord because it has reached that particular stage in its life cycle.

## The life cycle

The life cycle of an insect is important because it gives us a guide to when the insect is likely to damage the garden, and thus an indication of when and how to control it.

Most insects begin life as an egg and these may vary in size from quite tiny up to about 3 mm in diameter. Insects usually lay their eggs in groups, positioned differently according to the particular insect. Some insects, such as aphids, even hatch the egg inside the parent's body, but most insects lay their eggs on plant material.

The egg is very vulnerable to predators and so the parent goes to great lengths to protect her potential offspring. Some insects will roll the leaf over so as to form a protective tunnel in which to lay the eggs, while others will lay them on the underside of a leaf where they are less visible to predators. Others form a protective web of foliage, while the mantids cover their eggs in foamy material and the mealy bugs protect theirs with a mass of wax-like substance. As well as protecting the eggs from predators, these measures prevent them from drying out.

The attractive emperor gum moth caterpillars eat great lumps out of leaves, but is it rarely necessary to spray them — just hand pick and squash. (PHOTOGRAPH: NORTHSIDE PRODUCTIONS)

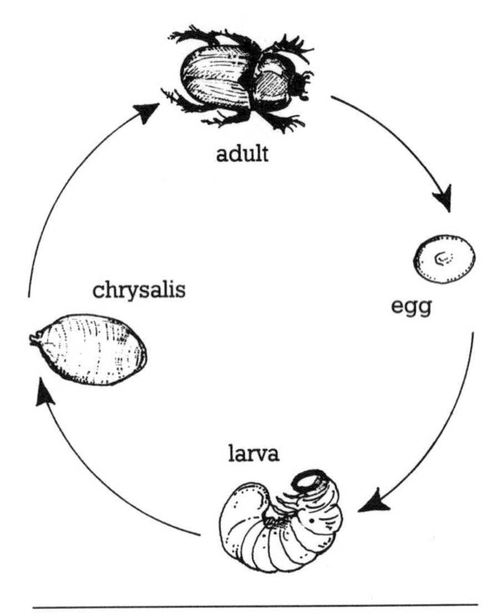


Life cycle of a chewing insect

The eggs of most of our common garden pests hatch out in a couple of weeks, but there are some that take only a few hours, while others can take months. The prevailing weather and environmental conditions will influence the time that the eggs take to incubate.

Eggs are not in themselves damaging to plants. It is usually at the next stage of the insect's life cycle when most damage occurs. This is particularly true of chewing insects which hatch as larvae, since it is at the larval stage that most feeding occurs. The act of laying the eggs can also cause plant damage. The cicada, for example, damages the twig as it lays its eggs. The longicorn beetle ringbarks a branch or twig and lays her eggs just above the ringed bark. This ringbarking causes the plant to build up the concentration of starch in the tissue near the damage and the resultant larvae are able to feed on a starch-enriched diet as they bore into the wood.

The gall wasp, which attacks citrus trees in the warmer regions of Australia, causes severe damage to the stems of the citrus tree in the process of laying its eggs. The affected plant parts swell up, restricting the transportation mechanism that carries plant nutrients and moisture to the extremities of the branches.

In most cases the adult insect lays its eggs on the plant material that will provide the larvae, or offspring, with food as soon as they are hatched out. The life cycle is illustrated at left.

Insects have a digestive tract, a heart, a respiratory system, muscular and nervous systems and reproductive organs. They are, in fact, more complex animals than we would at first imagine. I still remember my university days, when we were made to dissect a cockroach, as well as a range of other insects. All I had ever heard about cockroaches in my childhood came flooding back to me and it was one of the most repulsive tasks I have ever had to do.

The life cycle of an insect can be simple, like the one illustrated, or it can be quite complex, involving a number of body changes or metamorphoses.

## Plant damage

Insects cause damage to plants in a number of ways, and whilst their method of feeding is not the only source of damage, it provides us with some clues about the best method of control. The following list indicates the type of damage caused by some common pests.

1 Chewing insects: grasshoppers, caterpillars, beetles.
2 Sucking insects: aphids, thrips, mealy bugs, scale.
3 Borers: Acacia Stem Borers.
4 Root, stem and bulb attackers (mostly soil inhabiting): cutworms, grubs, bulb fly.
5 Tissue feeders: bean fly, Citrus Leaf Miner, Cineraria Leaf Miner.
6 Fruit eating: Macadamia Nut Borer, fruit fly, Codling Moth.
7 Gall forming: gall wasps.

## The chewing insects

The chewing insects have mouth parts known as 'mandibles' with which they are able to devour large quantities of foliage. It is possible to watch a caterpillar sitting on a leaf and see its head moving in a semicircular motion as it feeds voraciously on the edge of the leaf. When you remove the caterpillar, the damage that it has caused to the leaf has that same semicircular shape.

While most gardeners would suggest that chewing insects can be a curse in the garden, there has been a time when I welcomed one particular group of them into my own garden. It was quite a few years ago when I read some

German research in which it was reported that a building covered by a lush green climber or creeper would be cooler in summer by about 15°C. I made the mistake of telling my wife, Bev, about this and before you could say 'Jack Robinson', she'd gone out and bought some creepers to put on the house and garage. The creepers she bought were the *Parthenocissus quinquefolia* (Virginia Creeper).

Now it would be wrong to say that at the time I didn't like Virginia Creeper, because during my time in New Zealand I'd seen some lovely examples growing happily over buildings and, during the autumn, the colours of the foliage were quite beautiful. But, of course, that was in New Zealand where both the temperatures and humidity were much lower than in Brisbane. So we ended up with five of these Virginia Creepers and every one of them grew bigger and faster than any I had ever seen in my life before. The legendary triffid had nothing on them! They quickly covered the wall, climbed up under the eaves, and then proceeded to fill up the guttering. The novelty of pruning these things back every few weeks soon wore off, and what was even worse was that these plants started to show signs of an intelligence greater than I ever thought possible. They would surreptitiously send their tentacles through the ventilation holes in the eaves and then come out for air and light between the roof tiles, about three inches further up from the edge of the roof than my arm could reach from my position on the top step of the ladder.

But worse was to come. When I pulled them off the wall, the suction pads that held them to the surface were so firmly attached that the paint peeled off. Of course, everyone knows that the Virginia creeper is deciduous and every autumn those beautiful leaves fall from the vine. And where did they fall? Straight into the swimming pool, where they proceeded to stain the pool surface.

Help was at hand, however, in the form of the Hawk Moth Caterpillar. This giant of the insect kingdom has the ability to consume enormous amounts of foliage in a relatively short time, usually without detection, because it takes on the colour of the plant material it's feeding on. The arrival of these insects was heralded by their droppings on the ground below the creepers. In no time at all the majority of the leaves had been eaten and all I had to do was sweep the droppings into the garden beds.

Now, had I wanted to control these insects, I would have sprayed a contact or stomach poison, but of course killing them was the last thing on my mind as I welcomed their appearance in my garden each year. I'm not so benevolent when it comes to other insects — those caterpillars that get into my tomatoes, chew holes in my cabbages and eat out the kernel of my corn are not at all welcome.

Nasturtiums are widely grown as a companion plant for roses. These roses have never been sprayed and yet have not been attacked by aphids.

## The sucking insects

The sucking insects have very different mouth parts from the chewing insects in that they have a sharp pointed beak known as a 'proboscis'. They find their host plant and insert their proboscis into the sap-conducting tissue. It is really incorrect to call them sucking insects, as the fact is that very few, if any, need to suck at all. The pressure of sap within the plant, known as 'turgor' pressure, is at times quite powerful. While present, it helps to keep the leaves and stems firm and, at times, upright. As soon as the turgor pressure is lost, the plant leaves droop or wilt.

The sucking insect has long since learnt about this turgor pressure and knows that it only has to insert its proboscis into that sap stream and the turgor pressure will force the sap into its body. Unfortunately the so-called sucking insect has no control over the quantity of sap it ingests in this way and, indelicate as it may seem, the pressure is so great that the insect can only absorb a

This blue-tongue lizard has been doing a good job of insect control in Colin Campbell's garden in Maleny, Queensland.

certain amount of sap, while the remainder passes straight through. An interesting phenomenon occurs as the sap passes through the insect. Plant sap is a dilute mixture of sugars, starches and nutrients and is sugary and sticky. After it has passed through the insect, it is still sugary and sticky, but on the way through it changes its name from 'sap' to 'honeydew'.

The honeydew settles on the surrounding plant material and provides a suitable food source for a black fungal growth known as 'sooty mould'. This fungus is invariably associated with the presence of one or other of the so-called sucking insects and it cannot be eliminated from the plant until the insect that is producing the honeydew is removed.

Sucking insects will generally be found at or near the growing tip where the concentration of sap is greater. This is usually where the plant tissue is much softer and the proboscis is more easily inserted. Aphids are not known to try and put their little beaks into the hard trunk of a rose but they will put them into the growing points and buds.

Scale are not quite so particular, but they do prefer the softer material to the trunk. You will often find scale insects lined up along the central vein on the underside of a leaf.

There is quite a range of sucking insects, including aphids, thrips, scale and mealy bugs. Besides the damage they do to plants through their sucking habits, these pests may also be responsible for transmitting viral diseases from one plant to another. They are not known for their clean habits and they certainly never bother to disinfect their proboscises after visiting a diseased plant.

## The borers

Borers may affect various parts of the plant, but are usually associated with trunks, branches or stems. Poincianas and callistemons are prone to attack by borers and these usually attack the lower branches. The insect bores into the branch and the wide-awake gardener will notice a telltale trickle of sawdust down the trunk of the tree. The not so wide-awake gardener will probably only notice it when the foliage on that particular limb starts to die.

A number of Australian native shrubs, particularly wattles, are prone to attack by borers. Some species have a very short life in the warmer climates of Australia, simply because of borer damage. *Acacia macradenia* (Zig Zag Wattle) is one that is reasonably resistant and is worth a second look in borer-prone areas.

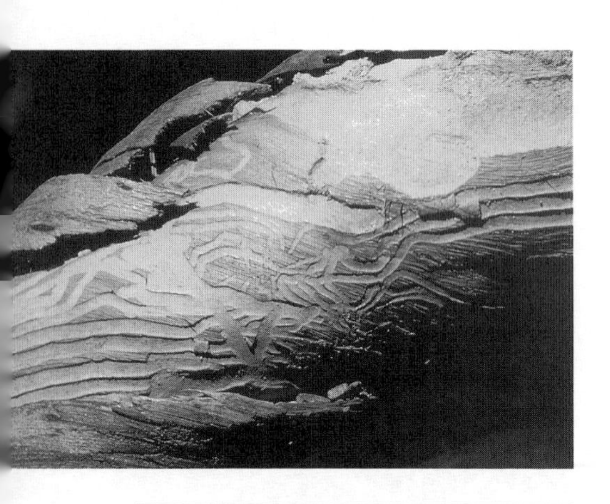

Jewel beetle larvae have caused this borer damage to a eucalypt trunk. The adult beetle often feeds on the leaves and flowers as well. (PHOTOGRAPH: NORTHSIDE PRODUCTIONS)

## Root, bulb and stem attackers

The roots of plants may be attacked by a range of soil-inhabiting insects. One of the most common in all regions is the cutworm, which just loves to destroy those lovely little seedlings. The beast in question works under cover of darkness and next morning, when you go out to give those treasured seedlings a drink, you'll find them all lying on the ground. A close inspection will reveal the fact that the stems have been chewed through just at or above ground level.

Lawns may also be attacked by root feeders, such as the white curl grub, while African black beetle larvae eat through the grass stems just at ground level.

Bulbs and tubers can often be attacked by a range of insects, some of which lay their eggs in the bulb. The resultant larvae eat out the tissue of the bulb or tuber and the bulb fails to grow.

## The tissue feeders

A number of insects lay their eggs just below the epidermis of the leaf. In many instances the only visible effect this has on the leaf is a tiny yellow spot at the point of entry. Again the egg hatches and the resultant larva tunnels its way through the leaf, feeding as it goes. In the case of the bean fly, the larva tunnels down the stem as it attempts to reach the ground in order to carry out the next stage of its life cycle. By the time it has got to ground level the larva has grown considerably and so the size of the tunnel has increased proportionately. The bean plant stem is so weakened that it is incapable of supporting the plant and it falls over.

Another insect with a similar feeding habit is the Citrus Leaf Miner, the larva of which tunnels around under the leaf epidermis, leaving a silvery to white trail as it goes.

While leaf miners are best known for the damage they cause to citrus trees, cinerarias and other flowering annuals, they also cause damage to weeds. Here we see a thistle which has been attacked by leaf miners, and this highlights the need for weed control as a means of controlling pests.

## The fruit eaters

The most common of these would have to be the fruit fly. This insect has similar habits to the leaf miners in that the female lays her eggs in the fruit and the larvae hatch out to fill the fruit with a mass of squirming grubs. Some fruit are more susceptible than others, but in general terms the softer the fruit, the more at risk it will be. Residents of the cooler regions such as Victoria and Tasmania should be eternally grateful that they do not have to contend with this pest.

Those regions do have their troublesome fruit eaters, though, in the form of the codling moth. This too results in finding a grub in the fruit or, as has been my experience, finding half a grub in a half-eaten apple. For those who do not wish to take any action to get rid of these pests, my only advice is to always eat susceptible fruit in the dark!

## Gall-forming insects

The Citrus Gall Wasp is one of the more common of this type. The parent lays its eggs on twigs, thorns or stems, and as the insect grows the swelling grows accordingly. There are no chemical controls for these insects.

Similar insects attack eucalyptus, casuarinas and certain other native plants. I remember seeing some of the native *Austromyrtus dulcis* (Midyim or Midgen Berry) that had been affected by a gall wasp and the effect was quite different from any other gall. These were round, thread-like galls about 2.5 centimetres long. These insects are dealt with in detail on page 21.

# Control

Control of these and other insects can be effected in a number of ways. Some can be controlled simply by removing them by hand, but while this is very effective it can be rather slow if there are a lot of insects. Some may be removed by squirting them with a strong jet of water from the hose, but chances are that a fair number of them will return as soon as they've dried themselves off a bit and can climb or fly back up to the plant.

Control by means of insecticides involves the use of one or other of the various chemical substances available to the gardener. These chemicals may be organic, extracted from plant material, or they may be man-made non-organic substances. Chemicals work in a number of different ways to protect plants from insect pests. They can generally be described as either systemic or non-systemic insecticides, and the latter group is further broken down into stomach poisons, contact poisons, general purpose poisons, systemic poisons and fumigants.

## Stomach poisons

These are substances that kill the insect when they are swallowed and taken into the digestive tract. They are used mainly to control the chewing insects. The chemical is applied onto the insect's food and as it bites off and swallows a quantity of that food, it also ingests the poison that covers the surface. These poisons are usually applied as a spray or dust.

A number of insects graze on the leaves of plants, causing damage such as this skeletonising. Applying a penetrative or contact insecticide is the most appropriate control method.

## Contact poisons

Contact poisons are insecticides which are able to kill the insect without being swallowed. Often the lethal agent is a gas which enters the spiracles of the insect and causes suffocation. Other compounds may affect the insect's nervous system. In olden times the most common contact poison used in gardens was the old Black Leaf 40, which was in reality a nicotine sulphate. While this was a most effective insecticide, it was extremely toxic to humans and was certainly an unsafe product to have around the house and garden. Pyrethrum is another contact insecticide and, of course, this is still in use today. Sulphur, and the various sulphur compounds, are also contact poisons which are effective against some of the mites.

This type of insecticide has limitations insofar as it tends to wash off the plant very readily and therefore does not offer much in the way of long-term protection. That lack of persistence, however, is recognised by the environmental movement as its greatest attribute.

Contact poisons are also effective against some of the sucking insects, such as aphids, but these insects must be actually present on the plant as they are being sprayed for them to have any effect. Repeat spraying will also be necessary as each new infestation arrives. The refined petroleum oils, such as White Oil, are also contact insecticides, but they suffocate the insect rather than poisoning it.

## General-purpose compounds

These include a wide range of insecticides, some of which are still in use . They usually act either as a contact or stomach poison, depending on how the insect encounters them. They will kill pests that eat them, are touched by them or breathe in the vapour.

These insecticides are sprayed or dusted over the surface of the plant parts to be protected. Most of them stay right where they are put and over a period of time they are broken down by rain, watering, light, or simply by the growing process of the plant itself. The persistence level of these varies considerably from a few days to a few weeks.

Some general-purpose insecticides actually penetrate the plant tissue. These are not to be confused with systemic insecticides described below. The penetrative insecticides are useful for tissue-feeding or mining insects which tend to operate just below the outer surface of the fruit or foliage. Bean fly and Citrus Leaf Miner are a case in point. The penetrative sprays can also be used for a wide range of other pests as well.

## Systemic poisons

This term describes substances that are absorbed by the host plant tissue and then carried to various parts of the plant by the transportation system. The process of absorption over various surfaces can, and does, present a few problems. Some plants are equipped with surfaces that actually repel these substances and therefore seriously reduce the insecticide's efficacy. Other plants are better adapted to absorb chemicals and thus the efficacy is enhanced. The addition of a non-ionic wetting agent will often improve the absorption of insecticides by plants with a water-repellent surface. Some gardeners have obtained beneficial results by adding a small quantity of Seasol to the insecticide mixture. In some instances the insecticide may be absorbed by the foliage, while in others the chemical is taken in through the roots.

The conditions for absorption either below or above ground are not always ideal. In the case of below-ground applications, high levels of organic matter can tie up the chemical and render it ineffective or, at best, slow down its progress into the plant's sap stream. In some cases this may mean that by the time the insecticide reaches the parts of the plant that are being attacked, the attacker has left and is undergoing the next stage of its life cycle.

In the case of foliage applications of a systemic insecticide, the problem of evaporation should be considered. All chemicals need to be diluted with water before they are sprayed over a plant. If the spraying is carried out during the hottest part of the day, the heat of the sun will quickly evaporate the water and, in some cases, a proportion of the chemical as well, which means the insecticide is no longer in a form which can be diffused into the plant tissue. Plants just do not have the ability to swallow up quantities of solid matter. Some

### Grease bands

This is an old home-made remedy that was used by gardeners many, many years ago. Mix eight parts of powdered resin with four parts of mineral turpentine and four parts of raw linseed oil. Add a small portion of honey and bring all of this slowly to the boil. Simmer for a few minutes and, while still warm, paint as a barrier around the trunk of trees. This barrier should be about 10 centimetres above the ground and will stop crawling insects from reaching the fruit and foliage.

### Lantana leaves

This weed has a use in the control of insects, both chewing and sucking, but it is said to be more effective in controlling aphids on contact. Boil 500 grams of leaves in 1 litre of water, strain, and use liquid when cool.

### Rhubarb spray

Rhubarb leaves contain oxalic acid, which is useful in the control of aphids. Boil about 5 kilograms of leaves in 1.5 litres of water for half to three-quarters of an hour. Strain off the liquid and add about 28 grams of Sunlight soap to it. Use this mixture in equal parts of water and spray over foliage of plants. Rhubarb spray is said to be non-toxic to bees.

Encourage native wasps, which eat many varieties of harmful insect larvae, to your garden with flowering plants. They are particularly attracted to daisies. (PHOTOGRAPH: ALLEN GILBERT)

### Methylated spirits

This is most effective in the control of minor infestations of mealy bugs. Simply dip a cotton bud into methylated spirits and wipe over the insects.

gardeners do not read the directions properly and as a result do not provide sufficient water to allow diffusion to take place, regardless of the temperature.

The systemic insecticide will either be transported by the xylem or phloem, parts of the transportation system of the plant. If there is no growth taking place because of cold temperatures or low light, then the movement of sap within the plant will be reduced. It should also be remembered that the xylem carries water from the root area to the remainder of the plant and it can only carry water and other dissolved substances, such as plant nutrients and absorbed pesticides, upwards. As the plant produces sugars and starches, through that wonderful process known as photosynthesis, these compounds are transported from the point of manufacture, which is usually the leaves, to the rest of the plant, including the roots. The phloem is the conducting tissue responsible for this and will carry these substances either upwards or downwards.

The time of year is important, and there is little value in spraying a systemic insecticide during a period of low growth. The chances are that the insects the chemical is designed to kill won't be about anyway, but in my experience there are a few 'trigger-happy' gardeners out there who spray insecticides around whether they are needed or not. The insects aren't quite as stupid and they will wait until periods of active growth, knowing full well that then, and only then, is there a plentiful supply of sugar-filled sap for them to feed on.

So, in order to be reasonably sure that the systemic insecticide is going to offer adequate protection, there are a few rules that should be followed.

- Apply the chemical during the late afternoon or on a cool day, to avoid the problem of evaporation of the water that will dissolve it.
- Always ensure that the plant is actively growing before application.
- Ensure that the correct dilution ratio is used to enable diffusion to occur.
- Only apply chemicals that are suitable for root absorption to the soil.
- Apply over as much of the plant surface as possible so as to ensure that both the phloem and xylem have the opportunity to carry the insecticide around the plant.
- Add a wetting agent or Seasol for hard-to-wet plants.

Systemic insecticides, more often than not, are more persistent than the non-systemic types but, because of the factors listed above, do not always have the quick knockdown characteristics of some of the non-systemic chemicals. This is not meant to imply that they do not have the ability to kill an insect on contact, because most insecticides can do that. If the insect is on the plant at the time of spraying, that generation will most likely die. If, however, the insects arrive a month after spraying, there may not be a sufficient concentration of insecticide in the sap of the affected part of the plant to kill the pests. In order to prevent that from happening, the wise gardener will know when insect pressure is likely to be at its greatest and will time the application for maximum effect.

Since systemic insecticides are taken into the sap stream, it should also be recognised that as the plant grows larger the amount of sap being transported around the plant will increase accordingly. This means that the insecticide is being progressively diluted and unless you adjust the amount used, it will eventually reach a level of concentration which is no longer effective. Most chemical recommendations take this factor into account and the repeat sprays allow for this. Make sure you follow those recommendations.

Because systemic insecticides tend to be longer lasting, when they are used on edible plants the chance of humans ingesting a quantity of insecticide is increased. In order to prevent that from occurring, it is essential that gardeners study carefully and adhere implicitly to the withholding period. We will discuss this at greater length later in this book.

It is important to recognise the feeding habits of insects so that the appropriate chemical may be used to control them. For instance, it is of little use applying a systemic insecticide to control caterpillars, while it is of about the same value to apply a stomach poison to control aphids.

If you consider the fact that a systemic insecticide will be located within the sap stream of the plant, it stands to reason that the greatest concentration of insecticide will be found in the veins of the leaf rather than the tissue. The chewing insect will start eating the foliage but will only ingest as much insecticide as can be found in the leaf veins. This may only provide enough insecticide to give the insect an upset tummy, but not enough to cause it to stop feeding or die.

If, however, a non-systemic insecticide is applied to that same leaf, there will be a coating of insecticide over the surface that the chewing insect is about to eat. If the insecticide has some contact action a certain amount of the chemical may be absorbed into the insect's body as it moves over the leaf. A further dose will be ingested as soon as the insect starts to chew on the foliage.

### Borax

Powdered borax and brown sugar is a fatal attractant for ants and cockroaches. Mix one part of powdered borax to five parts of brown sugar and place in small baits near areas where ants and cockroaches are a problem.

## Types of pesticides

In order to use pesticides responsibly, the gardener should learn something of each pesticide's action and use.

Pesticides may be classified as organochlorines, organophosphates, carbamates, inorganic insecticides or botanical or biological insecticides.

### Organochlorines

The organochlorines are sometimes referred to as 'chlorinated hydrocarbons' and while these were very effective in the control of insect pests, they were found to have a persistence in the environment which was considered to be unacceptable. This meant that they did not easily break down into their original components, or some other harmless compound. Whilst this aspect had benefits for the gardener and farmer, it raised serious concerns for the environmental movement when traces of these pesticides were found in food and animals. Some of our most effective insecticides were in this group, including DDT, dieldrin and lindane, but because of these concerns they are no longer available to the gardener or farmer.

### Organophosphates

The organophosphates are generally more toxic to mammals than the organochlorines, but they are usually much less persistent. This group of insecticides contains a number of more common systemic products, such as dimethoate and omethoate. Others of this group are penetrative, such as diazinon, maldison, and fenthion.

Carbamates are derivatives of carbamic acid and provide the gardener with a range of insecticides that are generally less toxic to us humans. As with all things, though, there are exceptions to the rule and not all carbamates are low in toxicity. We should never take the toxicity of any chemical for granted.

### Inorganic compounds

There are several inorganic compounds available for the control of garden pests. In earlier times lead and arsenic were in common use but, because of their extreme toxicity, they have been removed from sale. These days the various sulphur compounds such as sulphur dust, wettable sulphur and lime sulphur are the only insecticides available in this group. Sulphur is more commonly used as a fungicide and will be dealt with in that context.

### Fruit fly bait

There are a number of methods of constructing fruit fly baits, one of which is to drill some small holes in a clear soft-drink bottle, as described earlier. The holes should be near the top and just below these holes affix some yellow masking tape, since fruit flies are attracted to yellow. In the bottle place a protein attractant such as Vegemite, protein hydrolysate or even urine. Add to this, if you wish, an insecticide such as Malathion, and the fly will be attracted to the protein and killed by the insecticide.

You can also use repellants. One of these is a mixture of 1 litre of creosote, 1 litre of kerosene, and half a packet of moth balls. Mix these together well, wearing gloves and preferably a mask, as the smell is horrendous, and place in open tins tied throughout the fruit trees. This is said to keep fruit flies away, but take care not to walk into them when moving through the trees as it is not a good idea to spill this mixture on yourself.

### Plastic containers

Empty containers such as those which held ice cream, yoghurt or margarine are most useful in preventing cutworm damage to seedlings. Simply cut the bottom out of the containers and, at the time of planting out the seedlings, slip the container over the seedling and push it into the soil to a depth of about 5 millimetres. After about a fortnight you can lift the container off, as by then the seedling will be able to look after itself. Make sure you don't push the container into the ground any deeper than 5 millimetres, because if you do you will pull the plant out of the ground as you remove the container.

### Molasses

Molasses can be used to control caterpillars and grubs on a wide range of plants. Mix one tablespoon of molasses to 1 litre of water and spray over the target plants. Molasses is also very effective in the control of nematodes. Mix 4 litres of molasses in 12 litres of water and apply with a watering can, evenly over an area of 10 square metres. The molasses will not harm any plants that may be growing in the garden, but it may harm some worms.

### Organic compounds

A range of insecticides are derived from plant or vegetable material containing organic compounds that have proved effective in insect control. Pyrethrum is one such extract. It is taken from various members of the chrysanthemum family and is a contact insecticide. On its own, pyrethrum is not a terribly effective insecticide. It is relatively unstable and, although it has a fairly fast knockdown rate, it has little or no lasting effect. A synergist such as piperonyl butoxide is added to the pyrethrum to enhance its performance.

### Synthetic pyrethroids

Scientists have learned how to recreate these organic extracts in the laboratory and many of our pyrethrum sprays are now man-made. These are distinguished from the natural extracts by the term 'pyrethrins'. Pyrethrins have just the same ingredients and characteristics as natural pyrethrum, but they are usually much cheaper. They are also enhanced by the addition of piperonyl butoxide.

The synthetic pyrethroids were developed by scientists in an endeavour to capitalise on the safety and rapid knockdown characteristics of pyrethrum, while at the same time improving on the stability of the insecticide. Success came in the form of allethrin, then resmethrin, bioresmethrin, bioallethrin and finally phenothrin. These were all fifty times more potent in their ability to kill insects than the original pyrethrum, but they would break down very quickly in sunlight. This meant that they were of little value in horticulture, but they are used widely for domestic pest control and as fly sprays.

More recently, even more potent synthetic pyrethroids have been developed and these include a couple, fluvalinate and cyfluthrin, which have been released onto the home garden market. There is a downside to these recent developments in the area of synthetic pyrethroids though. Although potency has increased, so too has the toxicity of the chemical to humans. Also, the knockdown ability has decreased while insect resistance has increased.

Derris Dust is another organic insecticide containing the active ingredient rotenone, which is extracted from the roots of certain trees native to Indonesia and Malaysia. Eucalyptus and garlic are two other plants that are regularly used as ingredients in insecticides; later in this book we will deal with a range of home-made sprays that the gardener can make to control pests, using these and other common garden plants.

### Bacterial insecticides

*Bacillus thuringensis* is a bacterial insecticide which contains spores and is used to control caterpillars. The spores are from naturally occurring bacteria and are sprayed over the area to infect any caterpillars that may be present. The bacterial infection then kills the caterpillars. These bacteria are specific to caterpillars and are perfectly harmless to other animals, birds and insects.

Bud Worms can cause serious problems to flowers, particularly roses. The caterpillar either feeds off the tips of the bud or bores into its centre. Spraying with a non-systemic insecticide will effectively control this pest.

Ladybirds are a predator of aphids and some organic gardeners credit them with keeping mildew under control. The 28-Spotted Ladybird, however, is a pest, whereas the Common Ladybird, seen below, is a friend. (PHOTOGRAPH: NORTHSIDE PRODUCTIONS)

Leaf Curl is one of the most common fungal diseases to attack stone fruits. It causes distortion of the leaves and interferes with the production of sugars and starches by reducing photosynthesis.

Leaf blister galls caused by psyllids, common pests of many native plants. They are sucking insects and produce honeydew, on which black sooty mould forms. You can use a systemic insecticide to control this insect, but chemical control is unwarranted in most cases because of the size of the plants attacked. (PHOTOGRAPH: NORTHSIDE PRODUCTIONS)

This rose leaf is showing clear evidence of black spot, nitrogen deficiency (the yellowing) and one of the many viruses to which roses are prone. To treat black spot, spray with both a systemic and a non-systemic fungicide. (PHOTOGRAPH: ALLEN GILBERT)

An otherwise attractive lawn can be seriously damaged by one or other of the lawn grass pests. Here is an example caused by the Lawn Grass Caterpillar.

# Common pests and what to do about them

## Ants

I have a dreadful dilemma over ants in that my environmental instincts tell me that they are an integral part of the food chain and that any insect that has survived against the odds for so long deserves to live. On the other hand, my horticultural instincts tell me that ants can cause problems in potted plants by over-aerating the growing media. This can cause the mix to dry out, and roots that venture into the ants' tunnels lose contact with the potting mix, thereby decreasing their ability to absorb water and nutrients.

As well as that, ants have been known to 'farm' immobile sucking insects such as scale and mealy bugs. The ants will carry these insects to a suitable host plant, deposit them and wait for the sugary exudate, honeydew, which these insects produce — a highly prized food for the ant. When a gardener sees ants scurrying up and down the trunk of a tree, it is usually a safe bet that there will be one or other of the sucking insects producing honeydew.

### Chemical control

Chlorpyrifos, diazinon and pyrethrum are all registered for the control of ants. Spray or dust affected areas with a preparation containing one or other of these active ingredients. Where possible, apply directly to the nests.

African Black Beetle larva

### Non-chemical control

Place baits of five parts of icing sugar and one part of powdered borax in areas frequented by the insects. Boiling water poured down nests will also be reasonably effective, as will kerosene and pyrethrum.

## African Black Beetle

The adult beetles chew through the stems of the lawn grass plant just below ground level, causing the plant to wilt and fall over while the larvae eat the roots. As a result the affected area dies and the dead grass can be lifted off the ground like a carpet. Other plants that may be damaged by this pest include potatoes, tomatoes, sweet corn, cabbages, cauliflower, dahlias and petunias.

The adult African Black Beetle chews through the stems of plants just below ground level, causing the plant to wilt and fall over, while the larva eats the roots.

### Chemical control

Spray the area or the soil around susceptible plants with chlorpyrifos or fenamiphos. In New Zealand, chlorpyrifos granules are recommended.

The Lace Bug causes the leaves of azaleas to turn a dull, lustreless, brownish colour, sometimes with visible reddish brown spots on the underside. You will see similar symptoms when these plants are attacked by Two-Spotted Mites.

### Non-chemical control

There do not seem to be any reliable non-chemical controls for this pest, apart from maintaining adequate nutritional levels in the susceptible plants.

## Aphids

These pests congregate on new foliage, stems and flower buds where the concentration of sap is high and the soft tissue allows easy penetration. They come in a range of colours, ranging from yellow to pink, green, black and grey, and are usually 1–2 millimetres long. The sucking action causes distortion of the foliage and, in some cases, death. As mentioned earlier, aphids will transmit viral diseases from one plant to the next.

Aphids have been known to attack a wide range of plants, particularly at periods of rapid growth. Some are specific to a particular plant or species, but most plants are at risk.

### Chemical control

Systemic insecticides, such as dimethoate and omethoate, can be used for longer term control, whilst maldison, carbaryl, fluvalinate and pyrethrum will kill on contact. In New Zealand, acephate is also registered for the control of these pests.

### Non-chemical control

Hand removal or hosing with a strong jet of water will physically remove aphids. Soapy water and soap sprays are also effective.

## Azalea Lace Bug

These insects are found on the underside of the azalea leaf, where they suck the sap, causing the upper surface of the leaf to take on a dull, lustreless appearance. The leaf looks as if it has been sandblasted. Close inspection of the underside reveals reddish brown dots which are in fact the droppings of the lace bug. Control should be undertaken as soon as they are noticed as their activities cause severe disfigurement of the shrub. The lace bug will attack all members of the rhododendron family, including vireyas and azaleas as well as rhododendrons.

### Chemical control

Spraying with a systemic insecticide such as dimethoate or omethoate will be effective, as will fenthion.

### Non-chemical control

Pyrethrum sprayed under the leaves will provide some measure of protection.

## Azalea leaf miner

The adult of this insect deposits its eggs near the mid rib of the leaf and the caterpillars feed between the leaf surfaces. The damage first appears as thin white streaks which can only be seen from the underside of the leaf, but before long this develops into brown patches. The half-grown larvae migrate to the leaf tip and curl it over so as to provide some protection as it grows. In many instances the gardener notices the dead tip of the leaf before seeing the damage done earlier.

Azaleas are the main hosts of this pest but other members of the rhododendron family may also be at risk, although this is fairly rare.

Leaf miner damage first appears as thin white streaks on the underside of the leaf, then brown patches develop.

### Chemical control

Spray with dimethoate or diazinon about the middle of September.

### Non-chemical control

Remove curled over leaves as soon as they are seen and place them in a tightly tied plastic bag, which should be left in the sun to 'cook'. The infected leaves can then be placed in the rubbish bin and consigned to the refuse tip.

## Bean fly

This insect attacks all members of the bean family, including dwarf French beans and climbing beans; however snake beans and broad beans are resistant to it. The adult insect lays its egg in the leaves, causing tiny yellow spots to appear. The resultant larva then proceeds to tunnel its way down the stem in an effort to get to the ground, where the next stage of its life cycle is carried out. As a result of this tunnelling, the stem is weakened and this can cause the plant to fall over in the breeze, or when the crop is being picked. The yield could also be reduced. The problem of bean fly is confined to the warmer regions of Australia and is more prevalent in the warmer months.

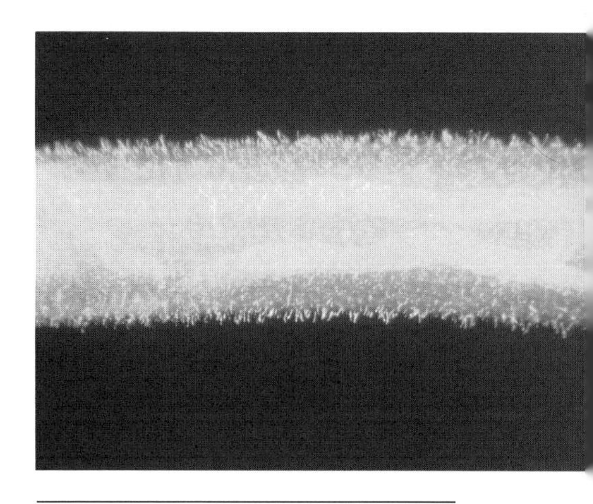

The bean fly lays its egg in the leaf. When the larva hatches out it tunnels its way down the stem toward the ground, feeding as it goes. The tunnelling effect can be clearly seen in this illustration.

### Chemical control

Spray the young plants with diazinon or dimethoate at an early stage.

### Non-chemical control

Apply Derris Dust to young plants. Hill the soil up around the stems to prevent breakage of the stem. Grow resistant varieties.

## Broad mite

Broad mites are very hard to see, but the damage they cause soon becomes obvious, showing up as a brown discolouration, often along the central vein. A wide range of plants are affected, particularly camellias and gardenias. Citrus trees and a range of other ornamentals are also likely to be attacked.

### Chemical control

Spraying with dicofol or wettable sulphur is usually effective. Dusting with dusting sulphur may also help to control these pests; however it is very difficult to place dust on the underside of the leaf where the insects are usually found. Fluvalinate is another effective miticide.

### Non-chemical control

Derris Dust is considered to be an effective deterrent to mites, but it is unlikely to eradicate mites already present.

Predatory mites are being used increasingly in nurseries against red spider mite and two-spotted mite. They are commercially available from Biocontrol, Warwick, Queensland, but tend to be most effective in controlled circumstances rather than domestic gardens. (PHOTOGRAPH: KNOXFIELD INSTITUTE OF HORTICULTURAL DEVELOPMENT)

## Borers

Borers attack a wide range of trees and shrubs. Many acacias are particularly susceptible, and borers are a very common cause of premature death in acacia species. Poinciana trees are also susceptible, but many other trees are at risk. Often the first indication that a gardener gets of their presence is the premature death of a limb or branch. A close inspection will reveal the hole, and tell-tale trails of sawdust may be visible down the trunk of the tree or on the ground below. In seriously affected cases, large areas of the tree may be dead. Pruning off those branches will show the borer holes.

Old yoghurt and other plastic containers with the bottoms cut out make a non-chemical, yet effective, control for cut worms when placed over young seedlings. They should be left in place for about two weeks, by which time the seedlings are no longer at risk.

### Chemical control

Use a hand sprayer with an adjustable nozzle to squirt maldison or carbaryl down the borer hole. Double-strength pyrethrum may also be used.

### Non-chemical control

Every wardrobe contains about three times as many wire coathangers as are necessary and one of these can be put to good use in the non-chemical control of borers. Cut a length of wire from an old coathanger and push it down the hole. Whichever way the borer is facing at the time will determine where the coathanger wire skewers it. Regardless of this fact, it is true that once the borer has been skewered on the end of a piece of coathanger wire, it immediately stops feeding.

## Bronze Orange Bug

The Bronze Orange Bug attacks a wide range of citrus and other fruit trees. This oval-shaped insect may be orange, green, brown, or even white. One thing that never changes, though, is the foul smell that it produces when disturbed. Bronze Orange Bugs appear in prolific numbers and cause fruit to fall because they have sucked the sap from the petiole which holds the fruit to the tree.

### Chemical control

Spraying with maldison, diazinon or dimethoate is effective, but dimethoate is not recommended because it can cause defoliation of Seville oranges, cumquats and Meyer lemons.

### Non-chemical control

These insects do not have a firm hold on the tree and can be shaken off quite readily. Shake them off into a tin containing a mixture of kerosene and water, but be prepared to put up with the smell for a while.

## Bud worms

These are caterpillars that bore into the flower bud or feed on the outside of the developing flower. They often start their life on the outside foliage and then work their way up to the flowering points. They can affect any flowering or fruiting plant, including roses, fruit trees and vegetables. The outer leaves of hearting lettuce and cabbage can also be affected by this insect.

### Chemical control

Spray unopened flowers with carbaryl at seven-day intervals, or with pyrethrum. It is important to note that the flowers must be at the unopened stage when you spray, as both these insecticides are damaging to bees. This will be dealt with later in the book.

### Non-chemical control

Dipel and Derris Dust may be effective in the control of these pests, and hand removal is also recommended.

The larvae of the Cabbage White Butterfly can cause serious problems in a range of vegetables and soft-foliaged plants. Spray with carbaryl, maldison, pyrethrum, Dipel or dust with Derris Dust. (PHOTOGRAPH: NORTHSIDE PRODUCTIONS)

## Cabbage white butterfly

The larvae of this insect can cause serious problems in a range of vegetables and soft-foliaged flowers. They feed on the outer and inner leaves and can cause quite a lot of damage.

### Chemical control

Carbaryl, either as a spray or dust, is effective, as is maldison and pyrethrum. These should be applied in accordance with the directions.

### Non-chemical control

Spraying with Dipel or dusting with Derris Dust will control this pest; however it should be done regularly and from an early stage of growth.

## Caterpillars

There is a wide range of caterpillars and, as we have already seen, these are chewing insects. The range of plants they attack is enormous, but soft-foliaged and soft-fruited species are most at risk.

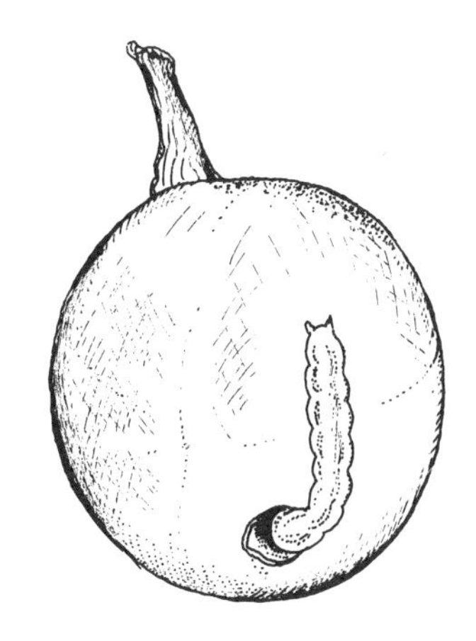

Many different caterpillars attack all sorts of plants, but they are especially fond of plants with soft foliage or fruit, such as tomatoes. Squashing the caterpillar and leaving it on the leaf is said to deter other caterpillars from calling.

### Chemical Control

Fenthion, maldison, carbaryl and fluvalinate are all effective in the control of chewing insects, being non-systemic. These will give a measure of control for a few days.

### Non-chemical control

Hand removal, Bug Juice, Derris Dust, Dipel, garlic sprays and pyrethrum are all effective caterpillar controls. It is also suggested that simply squashing the caterpillar between the thumb and forefinger and leaving it on the leaf will deter other caterpillars from calling.

## Christmas Beetle

I'm not sure why this beetle has been given this name, because it can certainly appear at times other than Christmas. As an adult it feeds on foliage and the damage has a characteristic saw-tooth shape. The larvae of this pest are the soil-inhabiting white curl grubs which can damage lawns and other plants by feeding on the roots. In the case of a lawn, as we have seen, the dead grass may be lifted off the ground like a carpet, while other plants that have been attacked may die in part or totally. The New Zealand grass grub, *Costelytra zealandrica*, is similar to this pest.

swelling caused by Citrus Gall Wasp

### Chemical Control

Drench the soil around the plants with chlorpyrifos or fenamiphos and spray lawns with similar chemicals. Most lawn grub sprays contain one or other of these ingredients or diazinon. Chlorpyrifos granules are recommended for control of the New Zealand grass grubs.

## Citrus Gall Wasp

This is a pest of warmer climates and it is very difficult to control. The adult lays its eggs in stems of citrus trees, causing the affected stem to swell and take on a greyish to white discolouration. The egg hatches out in the swelling, so it is important to effect control measures before hatching takes place.

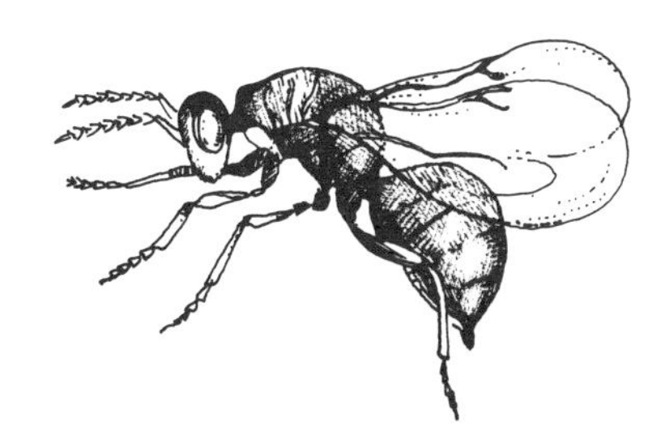

Citrus Gall Wasp (much enlarged)

Although the Citrus Gall Wasp is usually a pest of warmer climates, it has recently been found as far south as Victoria. The adult female lays its eggs in the stems of citrus trees, causing swelling and discolouration. The affected wood should be cut out and destroyed before the eggs hatch.

### Chemical control

There is no effective chemical control for this insect.

### Non-chemical control

At the first sign of any swelling, usually in late spring, affected wood should be cut out and destroyed immediately. If burning is not possible, the swellings

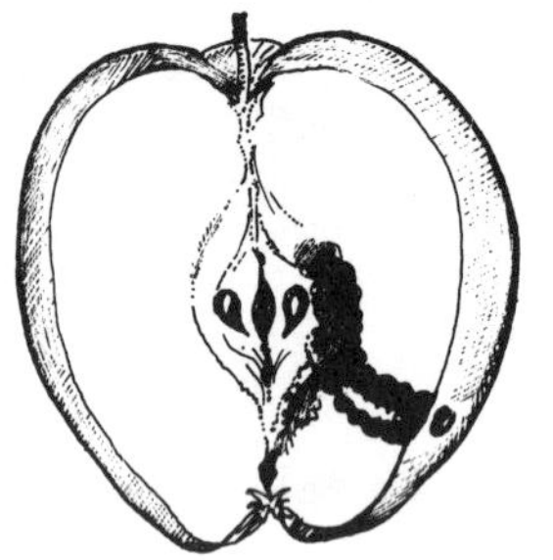


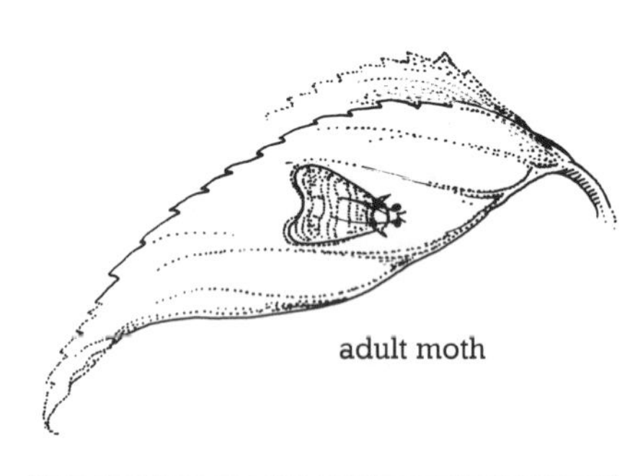


To control codling moth, spray with carbaryl or fenthion. For non-chemical control, remove and destroy affected fruit and any flaking bark where the insect may be lurking, and apply derris dust regularly from petal fall onwards. Spraying with dipel can also be effective.

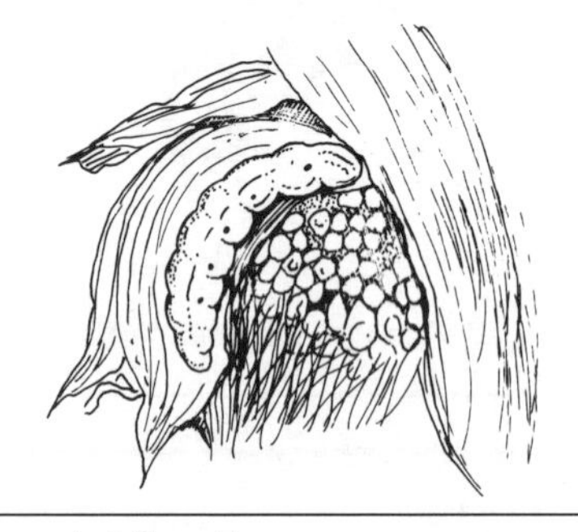

To control Corn Earworm, spray young corn with carbaryl, especially on the tip where the 'silk' forms, since that is the point of entry for the caterpillar. Derris dust and pyrethrum are also recommended.

should be placed in a plastic bag and tied tightly, then placed in the rubbish bin.

## Codling Moth

Codling Moth attacks apples, pears, quinces, crab apples, hawthorn fruit, and even walnuts and stone fruit, but it is most common in apples. The larva eats into the fruit core. The entrance hole is usually surrounded by webbing and the droppings that the insect has pushed out.

### Chemical control

Spraying with fenthion or carbaryl is recommended. If carbaryl is used it should be sprayed at petal fall, then repeated at fourteen-day intervals until December. Spraying at twenty-one-day intervals is needed from December on.

### Non-chemical control

Remove and destroy any affected fruit and any flaking bark, as the insect may reside in crevices on the trunk. Derris Dust applied at regular intervals from petal fall onwards is also said to be effective and spraying on a regular basis with Dipel may also be worthwhile.

## Corbie

This lawn pest is most common in the cooler regions such as Victoria and Tasmania. Its New Zealand relative is the Porina Caterpillar, *Wiseana cervinata*. It feeds on the lawn grass leaves under cover of darkness and large bare areas may appear on lawns overnight. It is most prevalent in autumn and winter.

### Chemical control

As soon as a bare patch appears, spray the lawn with carbaryl or one or other of the lawn grub sprays. The addition of some wetting agent will enhance the efficacy of carbaryl. New Zealand gardeners are advised to use chlorpyrifos granules to control this pest.

### Non-chemical control

There does not appear to be any non-chemical control for this pest, apart from encouraging birds.

## Corn Earworm

There is nothing worse than going out into the garden to pick a lovely fresh corn cob, only to find that the kernels have been cleaned up by the Corn Earworm. This caterpillar is one of the *Heliothis* family and gets into the cob as the kernels are forming. The same insect can attack beans and tomatoes. In the case of beans it will bore into the pod and then clean out the seeds.

### Chemical control

Spray the young corn with carbaryl, paying particular attention to the tip where the 'silk' forms as that is the point of entry for the caterpillar. The whole of the bean and tomato plant should be sprayed. In New Zealand, diazinon is registered for this purpose.

### Non-chemical control

Apply Derris Dust to the tip of corn and the stems of beans and tomatoes as a preventative measure. Spraying with pyrethrum is also recommended.

## Cutworms

Cutworms attack the young leaves and stems of newly planted seedlings and delicate young plants. It is most common in seedlings to find the stem eaten through and as a result the seedling simply falls over and dies. The caterpillar lives in the soil and generally comes out to feed at night.

To control cutworm, drench the ground around new seedlings with carbaryl or chlorpyrifos in the late afternoon or early evening, before the caterpillars emerge. It is important to mulch well around the plant and keep weeds down, too, as preventative measures.

### Chemical control

Drench the soil around new seedlings with carbaryl or chlorpyrifos in the late afternoon or early evening, before the caterpillars emerge.

### Non-chemical control

Consistent heavy mulching around new plants will help, as will sustained weed control, since the insect carries out a further stage of its development in weeds. It is also effective to wrap a strip of foil around the stems of seedlings, or to place a used plastic container, such as a yoghurt container, with the bottom cut out over the plant. Push the container about 5 millimetres into the ground.

To protect seedlings from cutworms, place a plastic container with the bottom cut out over the plant, pushing about 6 millimetres into the ground.

## Earwigs

These insects are often considered harmless, but they do in fact feed on both living and dead plant material. They can attack a range of vegetables and ornamentals; in particular lettuce, dahlias and chrysanthemums.

### Chemical control

Spray the area where they may be found with carbaryl.

### Non-chemical control

Setting traps is the only effective method of controlling earwigs without resorting to chemical methods.

Earwigs are often considered to be harmless, but they do feed on plant material and can attack a range of vegetables and ornamentals, especially lettuce, dahlias and chrysanthemums. Spray with carbaryl or set traps.

## Erinose Mite

This particularly troublesome insect attacks a range of plants but is mostly confined to *Euodia elleryana* and lychee trees. Lately, however, hibiscus in the Brisbane and Gold Coast regions have also been attacked. The symptoms of damage are a reddish brown blistering of the underside of the leaves. The blisters take on a velvety appearance.

### Chemical control

Control in *Euodia elleryana* is difficult and is probably not warranted, however it is important to embark on a spray programme in order to obtain optimum yields of lychees. This involves spraying the new growth with dimethoate and then doing at least four repeat sprayings at fortnightly intervals with one or other of the sulphur sprays. In the case of hibiscus, regular spraying with chlorpyrifos will help keep this pest under control.

### Non-chemical control

Unfortunately there do not appear to be any non-chemical control methods available at present. Considerable work has been done in an attempt to find a suitable predator, but so far this has been unsuccessful.

## Figleaf Beetle

This beetle feeds on the leaves of a number of ornamental and fruiting plants, including figs. The edible fig appears to be more susceptible than the large

Fruit fly is a major problem in the warmer parts of Australia. If you don't want to spray, follow the example of this gardener, who has covered her persimmons with brown paper bags.

ornamental figs. As the beetle feeds on the foliage it takes all the leaf tissue, leaving a skeleton of veins still intact.

### Chemical control

Spraying with carbaryl appears to be reasonably effective.

### Non-chemical control

The yellow eggs that are laid in late spring hatch out to become the larvae which cause the damage. Squashing these eggs by hand will, of course, prevent this damage from occurring.

## Fruit fly

This nasty pest is mainly confined to the warmer areas of Australia and one can only hope that it never becomes a problem further south. There are few fruits that are not susceptible. Avocados, bananas, citrus, figs, stone fruit, pip fruit, grapes, persimmons, loquats, passionfruit, feijoas, tomatoes and capsicums are just a few which are at risk. The female fruit fly can be seen standing on the fruit surface and she will often turn to face the on-coming gardener. She lays her eggs just below the surface and these hatch out, resulting in a seething mass of wriggling grubs.

### Chemical control

A number of chemicals are effective against fruit fly, but fenthion is the most effective. Dimethoate and omethoate will control fruit fly but there are a number of susceptible plants that will not tolerate being sprayed with either of those chemicals. Trichlorfon is also effective.

### Non-chemical controls

There are a number of suggested remedies that do not involve the use of chemicals, including hanging open tins containing a mixture of kerosene, moth balls and creosote. This evil-smelling mixture is reputed to act as a repellant.

A number of baits can be used. One of the most effective is to drill a series of holes, just large enough for the fruit fly to enter, around the top of a clear plastic soft-drink bottle. Because fruit fly are attracted to yellow, a strip of yellow tape just below the holes should be applied. Place some attractant in the bottom of the bottle and the fruit fly will enter the bottle but be unable to leave. A suitable attractant may be a mixture of water, Vegemite and Malathion, or even urine has been used successfully. A fruit fly trap which operates on a similar principle to the old fly paper is also being used. These tent-shaped traps are hung up in the trees and the fruit flies, being attracted to the sticky paper, fly onto the trap, where they are held and destroyed.

The fruit fly adult lays its egg just under the surface of the fruit, the egg hatches to become the larva and the fruit is rendered inedible. This peach has been attacked by fruit fly.

## Gladioli Thrips

These insects feed on the leaves and buds of plants. Affected leaves may have silvery streaks, while buds and newly emerging leaves may be distorted or deformed. Flowers often have a speckled appearance after attack by thrips. The adults usually attack the outer leaves and buds and the juveniles prefer the new leaves and flower buds. Thrips are quite small and are often hard to see with the naked eye.

### Chemical control

Dimethoate, omethoate, fenthion and disulfoton are all recommended for the control of thrips and, because of their systemic action, will give long-term control over these pests. Malathion and carbaryl may also be recommended, but these will kill only on contact and therefore the insect must be present at the time of spraying. New Zealand gardeners are advised to use fluvalinate or acephate to control these pests.

### Non-chemical control

Pyrethrum and pyrethrins, along with garlic sprays, will also achieve some measure of control of thrips, and as is the case with Malathion, the insects need to be on the plant at the time of spraying.

## Grape phylloxera

This aphid is a serious pest in grapes. It lives on the root system as well as the above-ground parts of the plant and causes fleshy yellow galls to appear. It is more prevalent in grapes on heavy soils than those on sandy soils, but all may be attacked at some stage. Over recent years the nursery industry has produced phylloxera-resistant root stock and this has reduced the problem enormously. Government regulations prohibit the movement of grape vines between certain areas and you should adhere to these regulations. If you are not sure, it is worthwhile finding out what regulations apply to your district.

### Chemical control

There are no chemical controls available for this pest.

### Non-chemical control

Adherence to the regulations mentioned earlier is the only method of controlling this pest.

## Grasshoppers

In the warmer regions these are a major problem, having an enormous appetite and causing great damage in an amazingly short time. To watch one of these insects devouring foliage is awesome. They have a continual feeding action and their mobility ensures that they are able to cover large areas if not brought under control.

Grasshoppers come in a range of sizes and colours, namely large, small, green and brown. The quantity of foliage they eat is relative to their size, but there usually seems to be more of the small ones at any given time.

### Chemical control

Controlling grasshoppers by chemical means is extremely difficult as very few of the commonly available chemicals appear to have much effect on them. The contact sprays, such as carbaryl and Malathion, do have a repellent effect and will cause the death of the small green grasshoppers. Penetrative insecticides such as diazinon and fenthion will also kill some of the smaller ones and send some of the larger ones off to the neighbour's backyard.

### Non-chemical control

Pyrethrum sprays, garlic and onion sprays and eucalyptus oils are all effective in repelling grasshoppers; however the level of efficacy will vary greatly. The most effective control of all is to get up early in the morning and sneak up on the unsuspecting grasshopper, catch it, then pull its head off. If this suggestion is a little repulsive, then try cutting them in half with scissors or secateurs.

A bait of borax and honey is also recommended for grasshopper control. The rationale behind this is that they will be attracted to the honey and the

**Spare those spiders**

**There is nothing worse than walking into a spider's web on a dark night. For hours afterwards there is that feeling of some monstrous spider crawling down your back. Despite this, we should strenuously resist the temptation to spray those spiders because they do carry out an important role in the biological control of insect pests. It has been suggested that if all the spiders were killed, life as we know it on Earth would cease to exist after eight short years.**

Locusts or grasshoppers can devour large quantities of foliage in a very short time. The most effective method of control is decapitation, either with your gloved hand or with secateurs. Grasshoppers tend to be rather sluggish in the early morning and this is the time for the gardener to carry out the dastardly deed!

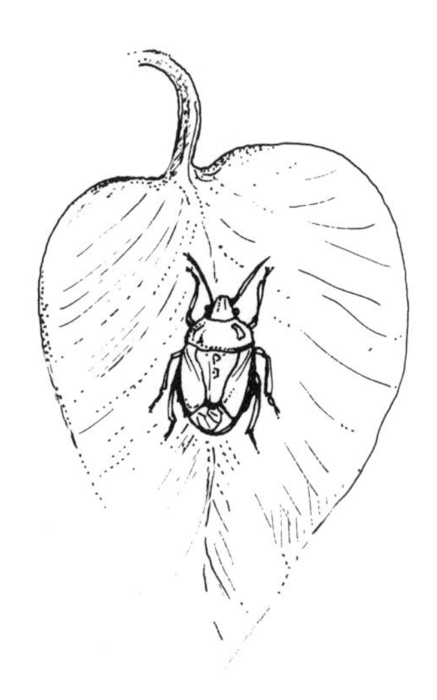

adult Green Vegetable Bug (actual size)

The Green Vegetable Bug attacks a wide range of plants. It sucks the sap from plants, causing distortion and yellowing of the leaves. Spraying with dimethoate, fenthion, omethoate, garlic, onion or pyrethrum will all help control this pest, but it is also important to remove plants after cropping and eradicate weeds.

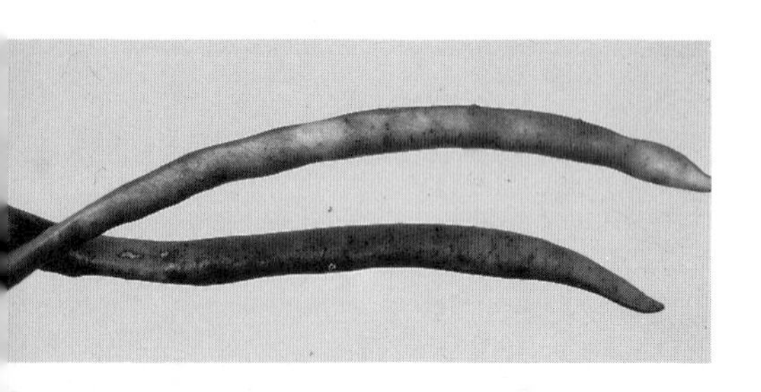

The Green Vegetable Bug attacks a wide range of plants but these bean pods are showing the classical symptoms of damage by this insect.

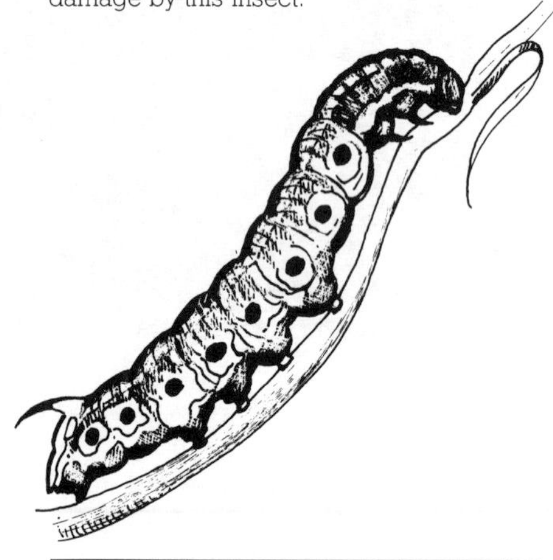

The giant Hawk Moth caterpillar (up to 75 millimetres long) is a voracious feeder. Pick off by hand or spray with a non-systemic chemical such as carbaryl or Malathion.

borax kills them. A fair number of baits will need to be spread about in order to control the large numbers of grasshoppers that might be present, however.

## Green Vegetable Bug

This pest attacks a wide range of plants, including pumpkins, capsicums, potatoes, spinach, oranges, peas, passionfruit and many other plants, and beans and tomatoes are under constant threat from it. It sucks the sap from the plants, causing distortion, yellowing of the leaves and serious distortion of young bean pods, which may also dry up. Tomatoes take on a mottled appearance in the affected leaves. The insects themselves come in a range of colours. The nymphs, or juveniles, may be yellow, red, orange, green or black, whilst the adults may be either green or black, but all have a distinctive shield shape.

### Chemical control

Fenthion, dimethoate and omethoate may all be sprayed on the plants to control this insect. Carbaryl and Malathion are also effective but as they do not have any systemic action, more regular spraying is needed.

### Non-chemical control

Removal of plants that have finished cropping and regular weed control will help minimise this problem. Garlic and onion sprays, along with pyrethrum and eucalyptus oils, will also be effective.

## Harlequin Bug

If it wasn't that this insect caused damage to plants, it would have to be one of the most welcome visitors to our gardens. The black and red, purple and orange colours make it a most attractive insect. Unfortunately it sucks the sap from young tissues and often causes the death of the plant, or at best the affected area. Fruit trees, rhubarb, melons, dahlias, tomatoes and hibiscus are all at risk from the Harlequin Bug.

### Chemical control

Systemic insecticides such as dimethoate and omethoate will give some measure of control; however, these should not be used on hibiscus as they can cause defoliation. Carbaryl and Malathion are also effective.

### Non-chemical control

Pyrethrum and derris dust may effect some measure of control, but neither is as effective as the chemical control methods. Hand removal by shaking the bug off into a tin of kerosene is probably the most effective non-chemical control method.

## Hawk Moth Caterpillar

This giant caterpillar is the larva of the Hawk Moth and is a voracious feeder. The caterpillar itself may be up to 75 millimetres long. It has a couple of dots on its back which look like eyes, but this is not so. It can be found on a wide range of plants, particularly those with soft foliage. Balsams, climbers and both ornamental and fruiting grapes are at risk.

### Chemical control

A non-systemic chemical, such as carbaryl or Malathion, sprayed over the foliage will help to keep this pest in check.

### Non-chemical control

Pyrethrum, derris dust, garlic and onion sprays are all effective in the control of Hawk Moth Caterpillar. Because of their size, they are easy to pick off by hand and this is by far the most effective non-chemical control, but they do have the ability to take on the colour of the plant material on which they are feeding, making them difficult to see.

The Hawk Moth Caterpillar has a voracious appetite and will consume huge quantities of leaf in a very short time. It will usually be found on the softer foliaged plants and, more often than not, it takes on the colour of the foliage upon which it is feeding.

## Hibiscus Beetle

This shiny black beetle feeds on the pollen of the hibiscus flower, causing the flowers to drop while in bud form. It is a particularly difficult insect to control and is probably the most troublesome of all the hibiscus pests. Regardless of the control methods used, hygiene is of paramount importance because good hibiscus management techniques will help break the life cycle of this beetle. Any unopened buds that fall to the ground must be picked up immediately, as the next stage of the beetle's life cycle takes place in the soil. Spent flowers should also be gathered up and some hibiscus growers even remove the older flowers from the bush at the end of each day. This prevents them falling to the ground and inhibits the progress of the Hibiscus Beetle. Some magnolias may also be affected by this pest.

### Chemical control

Carbaryl is registered for this pest but is only marginally effective, while some gardeners have achieved reasonable results through the use of diazinon. In both cases the spraying of buds should be carried out before any colour appears, as both insecticides are toxic to bees.

### Non-chemical control

Good garden hygiene, as mentioned earlier, is the only non-chemical method of controlling these pests.

## Lawn grubs

This term is used to describe a number of different insects, all of which cause serious damage to lawn grass plants. They include the African Black Beetle, the Lawn Army Worm, the Lawn Grass Caterpillar and the White Curl Grub. Damage may be caused either by subterranean feeding, in which case the roots of the lawn grass plant are eaten off and the lawn dies as a result, or surface feeding which causes browning of the grass and, in some cases, death.

The warm, wet, humid months are the most favourable for lawn grub activity, which suggests that these are more of a problem in tropical and subtropical regions than they are in the temperate zones. In order to find out if you have lawn grubs, lay a wet hessian bag on the grass overnight and when you lift it next morning, caterpillars and Army Worms will be on the surface of the lawn if they are in fact the cause of the damage. Subterranean feeders can easily be identified by grasping the dead lawn and trying to lift it off the ground. If the roots have been eaten off, the dead grass will lift off.

### Chemical control

A number of lawn grub preparations are on the market and these should be sprayed over the entire lawn at the first sign of damage. Ingredients of these preparations include chlorpyrifos, diazinon, fenamiphos and trichlorfon. In New Zealand chlorpyrifos granules may also be used.

Ladybirds are a predator of aphids and some organic gardeners credit them with keeping mildew under control. The 28-Spotted Ladybird, however, is a pest, whereas the Common Ladybird, seen above, is a friend.

### Non-chemical control

The red and blue Ichneumon Wasp is a predator of Lawn Grass Caterpillars and these are to be encouraged at every opportunity. Always make sure that any spraying is carried out late in the afternoon after these insects have gone away for the night. Birds are another predator of lawn grubs, so every effort needs to be made to encourage these into the garden. Spreading breadcrumbs and boiled rice over the lawn will bring in the ground-feeding birds.

Bandicoots are another of nature's lawn grub controllers and although they do disturb a lawn or garden in their quest for these subterranean feeders, they are an effective control method and should not be discouraged. Dipel, or *Bacillus thuringensis*, is another organic control of Lawn Grass Caterpillars.

## Leaf Blister Sawfly

This insect causes severe damage to members of the eucalyptus family and some other native species. The adult lays its egg beneath the surface of the leaf and the egg hatches out, causing a blister-like patch to develop wherever the larva feeds. In some cases the leaf may become transparent and you can see the larva beneath the surface when you hold the leaf up to the light. The foliage may be quite seriously disfigured; however it does not seem to cause serious harm or lasting damage to the affected tree.

### Chemical control

A systemic insecticide, such as dimethoate or omethoate, will be marginally effective, but it is probably not worth spraying because sooner or later, as the tree grows tall, spraying becomes impossible. Injection with one or other of these chemicals may be effective but it is unlikely that home gardeners will have the necessary expertise to carry out this task.

### Non-chemical control

There are no non-chemical methods of control for this pest, but control is usually unwarranted. Once the tree reaches about 4 metres high it seems to be less attractive to the Sawfly and is rarely, if ever, attacked.

Although many ladybirds are beneficial in the garden, since they eat aphids and other sap-sucking insects, there are also leaf-eating ladybirds which feed on the upper surfaces of leaves, causing them to turn brown and papery. You can generally tell the difference between friend and foe ladybirds by the number of spots — the beneficial kind usually have eighteen and the pests twenty-six to twenty-eight.

## Leaf-eating ladybirds

Most of us are aware of the beneficial ladybirds that prey on aphids and other sap-sucking insects, but as with all other insects there are both good and bad ladybirds. The good ladybirds usually have eighteen spots on their backs, whereas the bad ladybirds have twenty-six to twenty-eight spots. The damaging ladybird feeds on the upper surface of the leaf, causing it to take on a brown papery look. Plants that are most affected include the members of the cucurbit family, beans and tomatoes. Solanaceous plants, such as the potatoes, capsicums and a number of weeds, are also hosts to this pest.

### Chemical control

Spraying with non-systemic insecticides, such as Malathion and carbaryl, will be effective, as will dusting with vegetable dusts that include these insecticides.

### Non-chemical control

The only non-chemical controls that I'm aware of are hand removal or spraying with a garlic and onion spray.

## Macadamia Nut Borer

This caterpillar bores its way into the fruit of the macadamia nut tree but it is not confined to that species. Lychees and a number of other ornamentals are at risk. The insect penetrates the nut, eating out the kernel and leaving an empty shell. Because the kernel has been eaten, the nut usually falls to the ground long before maturity.

### Chemical control

The most effective control for the Macadamia Nut Borer is to spray with carbaryl as the fruit sets. Two or three sprays may be necessary in order to bring this pest under control.

### Non-chemical control

I am not aware of any non-chemical method of control, but gathering up affected nuts as soon as they fall would help to minimise the effect of the borers.

The Common Ladybird, with about eighteen spots, is a helpful predator.

## Mealy bugs

These sticky, white, furry-looking insects are sap suckers and their activity causes distortion and wilting of plants. They affect a wide range of plants, including palms, ferns, shrubs, fruit trees and fruit itself. Custard apples are a particular favourite because the mealy bugs are able to hide in the crevices of the fruit, but they can also be found down where the sheath emerges on palms. These hiding places are favoured because it is the only protection that they are able to afford themselves from predators, such as birds. Because they are very slow moving, they are quite vulnerable when out in the open, so they hide themselves away. Where there are no suitable hiding places, they will generally carry out their activities from the underside of the leaf or frond.

Ants are an indicator of the presence of Mealy Bugs and so is the black sooty mould that follows on the exudate (honeydew) they produce. Ants have been known to carry Mealy Bugs to a suitable host plant and then wait for them to produce the honeydew upon which the ants feed.

### Chemical control

Spraying with a systemic insecticide, such as dimethoate or omethoate, will control these pests and acephate is a recognised control in New Zealand.

### Non-chemical control

Pyrethrum or pyrethrins are effective and these are enhanced by the addition of one teaspoon of White Oil to 2 litres of spray mixture. If the infestations are relatively minor, it is possible to remove the insects by dipping a cotton bud into methylated spirits and wiping that over them. Obviously if the infestation is heavy, this method of control would be too slow. Badly infested ferns, African Violets and house plants of minimal value are probably best thrown away.

## Mites

Mites are in the same insect family as spiders, which may explain why many gardeners and gardening books refer to the Two-Spotted Mite as the Red Spider Mite, but of course they are not spiders as we know them. The Two-Spotted Mite is just one of the many mites that attack plants. The others include the Broad Mite; the Cyclamen Mite; and the Maori Mite, which attacks citrus trees.

Affected leaves often take on a silvery brown or grey appearance, the result of the feeding activity of the mite. They tend to suck the sap from the tissue, causing these colours to appear. Some viruses may be transmitted by particular mites while others secrete a substance which may stunt growth or cause a gall to form. Some of the mites are so tiny that they are barely visible to the naked eye and, in certain cases, are confused with the red droppings of the Azalea Lace Bug. In other instances they may cause fine yellowing or mottling of the leaves of fruit trees and, in the case of the Broad Mite, the central vein of the camellia leaf is the first place to show symptoms. The Maori Mite causes the skin of citrus fruits to take on a dull brownish colour.

Control of mites has become a major problem in the nursery industry, as many of the mites have built up resistance to commonly used chemicals in a relatively short time.

### Chemical control

Spraying with systemic insecticides, such as dimethoate and omethoate, will have limited success, as will the use of miticides, such as dicofol. If these have not been used in recent times, there is a chance that they will be more successful than if they have been regularly used over a long period. Fluvalinate is registered for the control of Two-Spotted Mites only and to date has shown good results.

### Non-chemical control

Spraying with sulphur sprays has been shown to be effective and mites do not seem to have built up a resistance to these insecticides, but be aware that there could be problems if this material is applied during hot weather because it causes leafburn. There will also be problems if it is applied to members of the cucurbit family. Biological predators are available and these are advertised in various gardening magazines.

## Orange Palm Dart

This moth lays its egg on palm fronds which it curls over to protect itself from predators. Under cover of this little tunnel, the larva feeds on the frond itself and is able to devour quite large quantities before being noticed. All palms are at risk; however the Alexandra (*Archontophoenix alexandrae*) and the Bangalow (*A. cunninghamiana*), along with the livistonias, seem to be most susceptible.

### Chemical control

Spraying with carbaryl or diazinon is effective, but, in order to make contact, the chemical must run down the 'tunnel' that the insect has created.

### Non-chemical control

Hand removal is the only suitable alternative to chemical control but it is relatively simple to carry out — all you need is a vigilant eye.

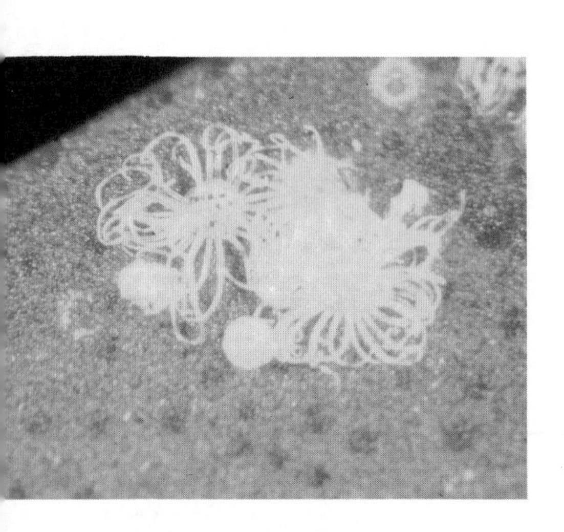

Psyllids (much magnified) on a melaleuca leaf. Found on many native plants, psyllids cause leaf blistering.

## Pear and Cherry Slug

Gardeners in the colder regions growing cherries, plums, apples and ornamental or fruit bearing pears will be well aware of these troublesome insects. They feed on the upper surface of the leaves until only a network of veins remain. Although called a 'slug', the insect is really a black, slimy caterpillar which can cause devastation to the fruit trees. Fruit yields will be affected because the plant is unable to produce sugars and starches through the process of photosynthesis.

### Chemical control

Spraying with a non-systemic insecticide such as carbaryl is probably the only registered control method.

### Non-chemical control

Spraying the insects off the foliage with a strong jet of water is possibly the only non-chemical control method available to the gardener, but it might also be worth trying one of the garlic and onion sprays. Pyrethrum sprays, applied as soon as the insect appears, can also be effective.

## Psyllids

These insects are found on many native species, particularly eucalypts and syzygiums. They cause a blistering of the leaf and, in the case of the syzygiums, it takes on a reddish colour. *Eucalyptus torelleana* is one that is commonly affected and in this case the psyllids produce masses of honeydew which literally drips onto the plants below. In some cases the psyllids will provide themselves with a protective covering which is known as a 'lerp'.

### Chemical control

Systemic insecticides, such as dimethoate and omethoate, are registered for use, as is the non-systemic Malathion, but the efficacy of any of these insecticides is questionable in many instances. New Zealand gardeners are often advised to use acephate as a control for these pests.

### Non-chemical control

There seems to be a case for not attempting control, either by chemical or non-chemical means. You can see syzygiums which have lived with psyllids for many, many years growing naturally in rainforests. The trees look perfectly healthy, so it is obvious that these insects are not doing them any harm.

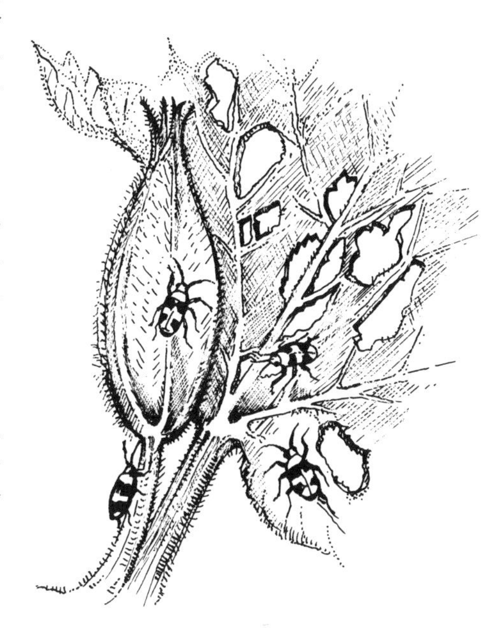

The native Pumpkin Beetle is yellowish orange with four black patches on its back. It attacks all members of the cucurbit family and can be controlled to some extent with carbaryl, pyrethrum or derris dust.

## Pumpkin Beetle

This native insect is yellowish to orange with four black spots on its back. It attacks all members of the cucubit family and occasionally may be found on other plants. The beetle feeds on foliage and flowers, and young plants are the most seriously affected. Some may even die as a result of this insect's feeding habits.

### Chemical control

Spraying with carbaryl at regular intervals will effect a measure of control, and so will dusting with one or other of the vegetable dusts that contain either carbaryl or Malathion.

### Non-chemical control

Derris Dust will certainly minimise the damage, as will pyrethrum or pyrethrin sprays.

The Pumpkin Beetle defoliates a number of plants, including pumpkins. All that is left after this insect has done its damage is a mass of skeletonised leaves.

## Monolepta beetle (Red-shouldered Leaf Beetle)

Rugby League fanatics will recognise this pest by its colouring, which resembles the colours of the Brisbane team, the Broncos. It is referred to by many gardeners as the 'Monolepta Beetle', which is in fact its correct name. It is more prevalent in warmer regions, particularly in avocado-growing areas. *Eucalyptus torelleana* is another of its favourite host plants. It is a gregarious insect, swarming in spring and late summer. The damage it causes can be to foliage, where it makes holes, or to flowers and fruit, both of which may be badly chewed. Besides avocados it will also attack most fruit, vegetables and ornamentals, including roses.

### Chemical control

This is extremely difficult as they are tough little characters to eliminate. Spraying with carbaryl or diazinon will have some effect but will not eradicate them completely.

### Non-chemical control

I am not aware of any non-chemical control methods, but it may be worth spraying with pyrethrum or garlic sprays.

## Scale insects

These may be found in pink, black, white and brown. They usually have a hard shell covering the body and this offers considerable protection from both sprays and predators. Their size and shape vary enormously; some are as large as a pin head, and others are shaped like a small pencil line. There is hardly a plant which is not susceptible to attack by scale insects and in many cases they are 'farmed' onto plants by ants. Ants will carry the scale, which is after all immobile, and deposit it on a suitable host plant. The honeydew that the scale produces provides a long-term food supply for the ant colony.

Scale insects are usually found along the central veins or stems. In this case they are on the central veins of a gardenia and, as is always the case, they have produced quantities of honeydew, upon which black sooty mould has formed. It will be necessary to remove the scale by spraying with White Oil and carbaryl before attempting to hose off the black sooty mould.

### Chemical control

The use of a suffocant spray is recommended, and petroleum oils are just that. White Oil is best mixed with carbaryl or Malathion before spraying. The White Oil suffocates the adult insect while the carbaryl or Malathion is effective against the 'crawler', which the insect often produces just before it dies. Systemic insecticides, such as dimethoate or omethoate, are also effective. In New Zealand, spray with an acephate plus White Oil mixture.

### Non-chemical control

It is possible to control scale insects with White Oil on its own, but this will have no effect on the 'crawler' and frequent spraying will be needed. Whenever you use White Oil, always remember that this operation should only be carried out in the cool of the evening as foliage-burning may occur in warm weather. Home-made White Oil is also effective and the recipe for this is given on page 60. For those who do not wish to spray with an insecticide and who have time on their hands, scrub the stems and branches where scale insects congregate with a mixture of washing soda and warm water. This will be most effective but, believe me, it is time consuming.

## Snails and slugs

These are not insects, strictly speaking, but they can certainly create havoc in the garden. Because they are nocturnal feeders, the damage they do doesn't become obvious until the next morning, but there's many a garden of seedlings or leafy plants that have been totally wiped out by these pests.

Shield bugs are welcome visitors to the garden since, as well as being very decorative, they feed on caterpillars. (PHOTOGRAPH: NORTHSIDE PRODUCTIONS)

Over-watering and under-watering can cause similar symptoms, but in the case of under-watering the tip of the affected leaf may be dry and brown, while over-watered plants may be soft and damp. These problems are most common in container grown plants. To test, push your forefinger into the potting mix as far as it will go. If the mix feels damp and cool leave well alone, but if it is warm and dry, water well.

Viral diseases are transmitted by aphids and other sucking insects and, once established in the plant, are impossible to control. Removal of the affected plant and its ultimate destruction is the only recognised method.

The sawfly larva does serious damage, particularly to eucalypts, causing the leaf to take on a brown, papery look. When the leaf is held up to the light, the larva can sometimes be seen inside the brown discolouration.

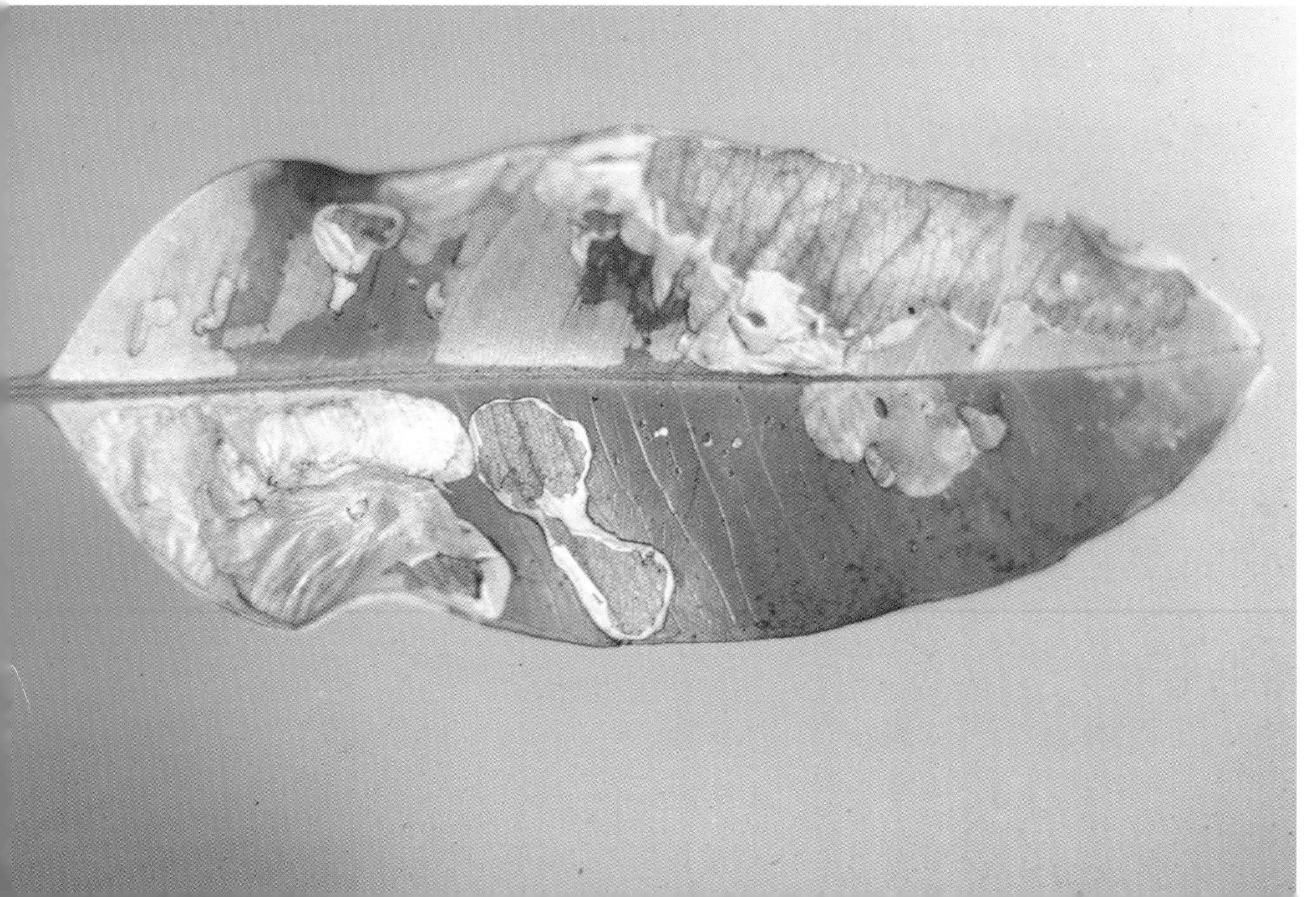

The stick case moth usually feeds on a wide range of small-leaved plants in the temperate and subtropical parts of Australia. It is not a serious problem, however, and can be controlled by hand picking. (PHOTOGRAPH: NORTHSIDE PRODUCTIONS)

Cup moth caterpillars can cause considerable damage to garden plants by skeletonising leaf surfaces at first, and eventually eating leaves right down to the midrib. They have a wide range of predators, however, including birds, lizards, frogs, wasps, viruses and fungi. (PHOTOGRAPH: NORTHSIDE PRODUCTIONS)

While the Harlequin Bug is a very attractive insect, it can cause quite serious problems. It sucks the sap on leaves and petioles, causing leaf and fruit drop.

There are more 'good' or beneficial insects in the garden than 'bad' or harmful ones; for example this mantid, which is about to polish off a fly. (PHOTOGRAPH: NORTHSIDE PRODUCTIONS)

### Chemical control

Snail baits containing metaldehyde or methiocarb are the usual methods of controlling these pests, but they can be harmful to domestic pets. Always place them in a position that's inaccessible to your pets and thus prevent problems.

### Non-chemical control

Sawdust sprinkled around plants is said to deter snails and slugs as they do not like to work their way across the gritty feel of the sawdust. Putting out baits containing stale beer is an alternative, as the snails apparently have a liking for the amber fluid. Gardeners with a liking for fresh beer, rather than stale, may have some difficulty in having any to spare for the snail baits. A size 8 shoe (or smaller or larger) placed firmly on top of the snail or slug is also a fairly effective control method, as are ducks, chooks and other birds. Lizards are predators of snails and slugs, too, so encourage them at every opportunity.

Thrip damage to a rose. Most plants are susceptible to damage by these tiny insects, but they can be controlled by chemical and organic means.

## Thrips

Thrips damage the leaf or flowers through the rasping and sucking action they use when feeding. Most plants are susceptible to damage by these tiny insects, which are barely visible to the naked eye. They often operate under the leaves of plants, causing a silvery grey appearance on the upper surface, particularly on older leaves. New foliage and flowers may be distorted or have colour break-down.

### Chemical control

Systemic insecticides, such as dimethoate and omethoate, and non-systemic insecticides, such as maldison and Mavrik are effective. Fenthion is also recommended, while in New Zealand, acephate is registered for their control.

### Non-chemical control

Dusting with Derris Dust or spraying with garlic sprays, pyrethrum or pyrethrins will help keep these pests under control.

## Webbing Caterpillar

This insect forms a web from the foliage of its host plant and sets about feeding in the protected environment of the web. Many native plants are susceptible to attack from this pest, but melaleucas, callistemons, and hakeas are particularly common targets. The presence of this insect is usually noticed when the tips of the branches suddenly die.

### Chemical control

Spraying with carbaryl or diazinon will bring this pest under control. In New Zealand, acephate is also recommended.

### Non-chemical control

There really isn't any need to spray these insects. Simply prune off any affected branches and the plant will probably be the better for the pruning anyway. This was often nature's way of pruning those native species which required it.

## Woolly Aphid

This is a problem in cool, moist areas where it feeds on the roots of plants as well as the woody above-ground parts and around fruit stalks. It is prevalent

### Insect-attracting plants

The following are some common plants that attract beneficial insects to the garden by providing nectar, an essential food source for predators and parasitoids. These then carry out a search and destroy mission on pests.

Alyssum (*Lobularia maritima*)
Borage (*Borago officianalis*)
Coriander (*Coriandum sativum*)
Fennel (*Foeniculum vulgare*)
Lucerne (*Alfalfa* species)
Queen Anne's Lace (*Ammi majus*)
Rue (*Ruta graveolens*)

A comprehensive list of insect-attracting plants, along with their cultural requirements and benefits, is given in an excellent organic gardening catalogue issued by Green Harvest, 52 Crystal Waters, MS16 Maleny, Qld 4552. The cost of this catalogue is just two current postage stamps.

## A short list of common pests and diseases and their controls

The following is a list of common problems that gardeners are likely to experience and the chemicals that are registered for their control. They are generally listed under the trade name of the product and this is the name by which garden retailers should know them. Before using any product, though, check to see that the pest or disease you wish to control is in fact listed on the label and note any cautionary statements relating to its use.

**Aphids**
Pyrethrum- or pyrethrin-based sprays; Carbaryl; Malathion; Rogor; Folimat; Mavrik (New Zealand: Target; Orthene)

**Black spot on roses**
Triforine; Rose Spray; Rose Black Spot and Insect Killer; Black Spot and Insect Killer (New Zealand: Shield; Bravo)

**Chewing insects**
Carbaryl; Malathion; pyrethrum- or pyrethrin-based sprays; Derris Dust; Mavrik; Caterpillar Killer; dipel (New Zealand: Orthene; Carbaryl)

**Fruit Fly**
Lebaycid; Rogor; Folimat.

**Fungal diseases**
Copper spray; Bordeaux mixture; sulphur spray; lawn and garden fungicide; Zineb (New Zealand: Bravo)

**Lawn pests**
Grubkil; Chlorban; Lawn Grub and Insect Killer; Lawn Beetle Killer (New Zealand: Soil Insect Granules)

**Leaf Miners**
Diazamin; Lebaycid (New Zealand: Orthene)

**Scale**
White Oil mixed with either Carbaryl, Malathion or Rogor.

during the warmer months and affects apples, hawthorn, cotoneasters and, occasionally, pears. The plant's lateral growth is distorted and buds may be destroyed. The insect is purplish brown in colour, but covers itself with a waxy white thread.

### Chemical control

Systemic insecticides, such as dimethoate or omethoate, can be used but it is important to spray the chemical thoroughly into any cracks, crevices or junctions as these are hiding places for the Woolly Aphid. New Zealand gardeners can use either diazinon or acephate to control this insect.

### Non-chemical control

Pyrethrum can be sprayed but will only kill on contact. Keeping the garden weed free and gathering up dead wood will also help to minimise the effects of this insect.

## Others

There are, of course, many other pests that may attack your garden. If you find one that you are unsure of, always seek advice before spraying with an insecticide. Your local Department of Agriculture or Primary Industries can often help with this, or try a nursery that has suitably qualified staff.

ZERO WEEDSPRAY is an extremely effective, easy and economical way to kill weeds and grasses in the home garden.
ZERO is not active in the soil (non residual) and biodegrades into natural products.

DATE OF MANUFACTURE:
Ⓑ

**HOW TO USE:**
DO NOT disturb treated weeds for 2 weeks.
DO NOT spray weeds if rain is likely within 6 hours.
Use Zero® Ready to Use Weedspray on weeds and grasses when they are actively growing and not under any stress. Visible symptoms of use will appear in 7-10 days.
Use Zero® Ready to Use Weedspray in rockeries, for spot treatment in lawns, potted plants, paths/driveways and for lawn edges.

| WEEDS CONTROLLED | HOW TO APPLY |
| --- | --- |
| Annual Ryegrass, Bamboo (NSW & QLD only), Barnyard Grass, Blady Grass, Brome Grass, Buffalo Grass, Couch Grass, Guinea Grass, Kikuyu, Lantana (NSW & QLD only), Nut Grass, Onion Weed (all states except SA), Oxalis species (Soursob in SA), Paspalum, Prairie Grass, Rhodes Grass, Sorrel, St. John's Wort, Winter Grass. | Apply to actively growing plants when most have reached the early head stage. Spray weed leaf area. Repeat treatments for perennial species. Apply to Onion Weed, Oxalis species (Soursob in SA) when flowering. |

**NOT TO BE USED FOR ANY PURPOSE, OR IN ANY MANNER, CONTRARY TO THIS LABEL UNLESS AUTHORISED.**
**CAUTION:** DO NOT allow spray to contact or drift onto plants you do not want killed.
DO hose off immediately if accidentally sprayed.
DO NOT allow chemical containers or spray to get into drains, sewers, streams or ponds.
**STORAGE AND DISPOSAL:** Store in the closed, original container in a cool dry place out of the reach of children. Do not store in direct sunlight. Dispose of empty container by wrapping in paper, placing in a plastic bag and putting in garbage.
**SAFETY DIRECTIONS:** Avoid contact with eyes and skin.
Wash hands after use.
**FIRST AID:** If poisoning occurs contact a doctor or Poisons Information Centre. If swallowed, DO NOT induce vomiting. Give a glass of water.
For further product information, write to:
Zero Consultant
P.O. Box 250 ERMINGTON NSW 2115
® Registered Trademark of R&C Products Pty. Ltd.
Made by Reckitt & Colman Products
33 Hope Street, Ermington NSW 2115 AUSTRALIA
19695

The information on garden chemical labels is there for your protection. *Always* read it carefully and *never* store chemicals in unlabelled containers. ('ZERO RTU' LABEL COURTESY OF RECKITT & COLMAN PRODUCTS)

# Plant diseases

In its broadest sense, disease in a plant is any alteration that interferes with the normal structure, functions, or economic value of that plant. In a garden this may mean that the plant will become deformed, wilt, fail to produce flowers or fruit, or, at worst, die.

Plant pathology is a relatively new science, only a few decades old in fact, but that doesn't mean that people have only had to live with plant diseases for that length of time. Amongst the earliest written records, unmistakable references to blight, mildew and plagues show us that plant diseases have in fact shadowed the agricultural path of humanity since people first scratched the soil with a pointed stick and then planted seeds.

In early Roman times a rust god, 'Robigus', was annually honoured by farmers in the hope that this would prevent outbreaks of the disease caused by rust. History has recorded many serious sudden and widespread outbreaks of plant disease. These can be compared to an outbreak of disease in humans or animals.

The Great Potato Blight of 1845 was one of these serious epidemics. It devastated millions of acres of potatoes in Europe, the USA and Canada, and it has been suggested that it spurred the world's scientists into studying plant diseases. A German, Heinrich Anton de Bary, stared at a dying potato leaf through his primitive microscope and after a time noticed that the green leaf cells were in the clutches of sinuous and pallid fibres which he identified as a fungus. He proved that this fungus was the cause of the potato blight and, as a result of his work, the way was cleared for Millardet, a French scientist, to provide an effective weapon against any recurrence of the blight.

At this stage it is probably worth recounting the story of the discovery of one of our oldest fungicides, Bordeaux Mixture. According to tradition, a farmer in Medoc, France, was anxious to stop passersby stealing his grapes. In order to deter the grapenappers, he decided to put some lime in water and throw it over the grapes.

Having done this, he was a little disappointed in that the grapes still didn't look repulsive enough and so he went back to his shed and added some bluestone or copper sulphate to the lime and water mixture. He splattered this mixture over the vines, then put up a sign at the end of the row stating that the grapes had been sprayed with some poisonous mixture, suggesting that anyone who ate the grapes would feel positively ill at best or, at worst, die.

It is not known whether Millardet was one of the grape thieves, or whether he was called in. What *is* known, though, is that he has been credited with the discovery that the grapes thus treated were the only ones not affected by powdery mildew.

Powdery mildew forms a grey, ash-like substance on the upper surface of leaves, as can be seen on these pumpkins.

A seedling nursery like this will take precautions to prevent an outbreak of pythium, commonly called damping off. Affected plants rot off at ground level, but spraying with a fungicide such as Fungarid will prevent this disease.

Powdery mildew was the scourge of the Bordeaux vintners and so Bordeaux Mixture was recognised as being their saviour. That same mixture is still being used in gardens and vineyards today.

Over the years there has been a number of other serious diseases, besides potato blights, that have affected large areas of horticulture. These include the Dutch elm disease, still one of the greatest threats imaginable to those majestic and wonderful trees. This has even found its way to the southern hemisphere and we must all be aware of the potential threat that it poses to Australian elms.

Fire blight of apples and pears can seriously affect the production of these crops. It has not yet been found in Australia and hopefully never will be, but again we must be vigilant, and this is why plant quarantine is so important.

Banana growers are aware of the devastation that Panama disease can cause. The whole of our banana industry could quite easily be devastated if this disease ever became uncontrollable. Despite the fact that Australia is an island, we still have our share of serious plant diseases and the risk of importing even more.

Plant diseases can be classified as those caused by viruses, those caused by parasitic animals or those caused by pathogenic plants. Other diseases are caused by non-living factors.

## Physiogenic plant diseases

A group of plant diseases is described as 'physiogenic', meaning caused by factors other than living organisms. They are abiotic or non-parasitic, and may be caused by such environmental disturbances or deficiencies as an imbalance of soil nutrients; excesses of soluble salts in the soil; unfavourable water, air and light relationships; unfavourable temperatures; harmful chemicals; or mechanical injuries.

### Soil nutrients

Deficiencies in one or more of the essential plant foods can result in characteristic plant symptoms. The most common deficiencies are of nitrogen and phosphorus, but potassium, and all the other nutrients, are also able to cause plant deficiency symptoms.

Deficiency symptoms are often confusing to the untrained eye and, more often than not, multiple deficiencies produce masked symptoms. If you watch your plants carefully, however, you will notice the deficiency symptoms as soon as they appear, and the more you know about them, the better you will be able to diagnose and correct them.

#### Nitrogen

Nitrogen deficiency always shows up in the older leaves first. They will gradually turn pale green, then progress to yellow as the deficiency becomes more pronounced. As nitrogen is the cornerstone of plant growth, a deficiency will cause a reduction in the rate of growth.

The most common soil nutrient deficiencies are of nitrogen and phosphorus, but plants can show symptoms of deficiency in all the other nutrients; for example, lack of sufficient boron causes fruit drop.

#### Phosphorus

This might be noticed when the plant flowers less, produces fewer seeds or is generally stunted. Affected plants look spindly and the leaves may take on a blue–green or purple discolouration before gradually yellowing. Root formation will also be inhibited.

### Potassium

Potassium deficiency also shows up in the older leaves first, with those leaves becoming a dull grey-green and yellowing in spots on the leaf margins and tips. The spots expand and scorching appears.

### Magnesium

Magnesium, when deficient, causes the leaves to take on a mottled, patchy, yellowed appearance. In some cases dead spots appear in yellow areas, while sometimes in others a brilliant colouring appears around the margins.

### Molybdenum

Molybdenum deficiency causes a mottling over the whole of the leaf, with some leaves becoming paler, and the leaves and stems may be severely distorted. A disorder in brassica crops, known as 'Whip Tail', is caused by a molybdenum deficiency.

### Other trace elements

Deficiencies in other trace elements may also cause symptoms. In the case of the acid-loving shrubs, iron may be deficient if the soil pH is too high, and this will manifest itself by a yellowing of the new leaves between the veins. In serious cases the leaves will go completely white.

Zinc deficiency causes a mottling of both the younger and older leaves, but the younger leaves display the most dramatic symptoms. Fruit trees, particularly citrus, are prone to this problem.

Club root is a disorder of cruciferous plants, such as cauliflowers and cabbages, and is more prevalent in soils low in the trace element, boron.

All of these disorders can be put to rights by applying the appropriate fertiliser containing the nutrients that are deficient; however, further disorders can be caused by excessive amounts of these nutrients. Always follow the manufacturer's directions when applying fertiliser nutrients.

The quality of potting mix varies enormously and this is well illustrated above. The two containers were filled with different potting mixes and the results are evident. Always buy a premium grade standardsmark potting mix to ensure good results.

## Excess salts

Excessive amounts of soluble salts in the surrounding soil can cause serious problems in plants. A physiological disorder known as 'plasmolysis' occurs, resulting in a loss of sap from the plant. If the excess of salt is only slight, the leaf margins might scorch or the plant might drop leaves, but in the case of high concentrations, the plant usually dies.

Use fertilisers with a low salt index and again follow the manufacturer's directions. Leaching the soluble salts away from the root zone will also minimise damage.

## Unfavourable water, air and light relationships

Plants need all of the above in order to make satisfactory growth and all must be in the right balance for the needs of the particular plant. More often than not, lack of light causes leggy growth as plants struggle to reach what available light there is, or failure to flower. Plants locked up in houses while their owners are away for long periods may suffer from a lack of humidity.

All plants need moisture in order to survive and too little or too much can have an equally disastrous effect. Too little moisture means that sap production is minimised and does not provide for the necessary movement of nutrients and other essential items throughout the plant system. On the other hand, excess water around the plant roots inhibits the movement of oxygen from the soil into the roots and, as a result, the plant suffocates.

When it comes to fertiliser applications, more is not necessarily better! In this instance serious damage has been caused to these palms by excessive amounts of fertiliser. These are classical symptoms of salt damage.

### Unfavourable temperatures

All plants have an optimum temperature range and this is evidenced when one tries to grow a temperate plant in a tropical climate or vice versa, resulting in the death of the plant. As well as temperature, humidity is also a factor and when either of these is incorrect, plant growth will suffer or the plant will die.

### Chemical and mechanical injuries

Some chemicals are 'phytotoxic'. In other words, they have the ability to damage the foliage, flowers, or fruit of plants. An example of this is dimethoate, which quite often causes defoliation of certain citrus trees and hibiscus. White Oil, if sprayed on a hot day, can also burn plant foliage.

I remember encountering a nursery with a serious plant damage problem, which was caused by the exhaust gases of cars as they turned around in the car park. The plants in the line of fire from the exhaust gases suffered extreme stress and many of them died. A physical barrier between the cars and the plants overcame the problem.

Closer to home, fumes from kerosene heaters can cause quite serious problems for any nearby indoor plants.

Mechanical injuries are caused by humans or animals. In the case of humans, it is generally as a result of improper use of garden tools such as lawnmowers, weed-eaters, spades, shovels and forks. Care in their use will prevent unnecessary disease problems which can follow injuries.

## Diseases caused by living organisms

### Viruses

Viruses are so small that an electron microscope must be used to see them. These organisms live inside the plant cell and this factor is important when considering their control. Any chemical which kills the virus will also kill the plant cell and hence the plant.

Many animal diseases are caused by viruses, while in plants, diseases such as mosaic diseases of tomatoes, potatoes and cucumbers, the yellows of peach and aster, and curly top of sugarbeets are viral.

Many plants can be affected with viruses and yet show no visible signs of the disease. These plants are known as 'carriers', simply because they carry the disease without succumbing. The importance of these carriers is easily understood. In a garden, certain weeds can carry the disease and act as a source of infection for nearby garden plants. In other words, the disease is transmitted from the weed to the wanted plant. Obviously weed control can, therefore, also act as a form of disease control and this is a very important reason for keeping the garden weed free.

A more difficult situation arises when certain plants used for seed or flowers are carriers and their introduction into the garden may result in severe viral diseases occurring amongst the other wanted species.

#### Transmission

For transmission it is necessary for the virus to come in contact with the host plant. This is most successfully achieved by insects that have a sucking habit. Aphids, scale, mealy bugs and thrips are just some of the transmission vectors. The long stylet, the part of the proboscis of the sucking insect that allows the insect to suck up the sap, penetrates the leaf tissue. As well as sucking up the plant sap, the insect also sucks up the virus from the diseased plant. It then moves to another plant and inserts its stylet into that plant's tissue. In doing so it releases a quantity of saliva and this saliva contains the virus which in turn infects the new plant. It is also possible for certain viruses to be transmitted in the seed. In other words, seeds of plants may have been infected with the virus and from then on all successive sowings of those seeds will produce virus-infected plants.

### Control

As already mentioned, the fact that viruses live inside plant cells makes them very difficult to get at and control since the chemicals that will kill the virus will also kill the plant cell. In order to control these diseases, it is necessary to control the insect vectors. In the previous chapter we discussed methods of controlling thrips, aphids, mealy bugs and other sucking insects and the use of appropriate insecticides is an integral part of viral disease control.

Garden hygiene is of paramount importance and the cleaning up of weeds and other diseased plant material will help keep viral diseases in check. On no account should any of these materials be put into a compost heap, as this may encourage their proliferation. In commercial plant growing, seed treatment and selection is used as a means of reducing the likelihood of diseases caused by seed-borne viruses.

The quarantine regulations that are sometimes quite frustrating after an overseas holiday are of vital importance to the future of horticulture and agriculture in this country. We should never complain about the rigidity with which the quarantine officers perform their duties at international airports, since their vigilance is necessary to our agricultural and horticultural industries.

## Diseases caused by parasitic or predatory animals

These parasitic or predatory animals include such things as nematodes, snails and slugs, insects, rodents and birds. They cause some damage to a particular part of the plant, providing an entry point for fungal, bacterial or viral infection. In order to control these diseases, it is necessary to control the creature causing them.

### Nematodes

Probably the most common and difficult to control is the nematode — an extremely small, roughly cylindrical worm which, although thin, may grow to 2 millimetres long. Such a tiny insect is usually invisible to the naked eye and there may be countless thousands of them in each garden bed. Sandy soils are often more susceptible to nematodes; however there are heavier soils which are also inhabited by this pest.

The nematode attacks the plants, more often than not confining itself to the roots, but there are examples of leaf nematodes. They can cause considerable damage to root and other crops. Potatoes and carrots are particularly susceptible. The nematode bores into the potato or carrot and immediately injects a substance which causes a gall or knot to form. These can make the potato tubers and carrot roots unsightly and, in extremely bad cases, inedible.

The root knot nematode is also a major problem for domestic tomato growers. Again, it burrows its way into the root system and the gall forms in the same manner as in potatoes and carrots, but unfortunately these galls or swellings inhibit the uptake of water and nutrients in tomatoes. In a very short time the leaves start to wilt and eventually the plant dies, usually long before fruit has been produced.

A similar fate occurs with the fibrous rooted plants, such as cabbage and cauliflower, although the results are not quite so dramatic and a yield of some sort, although a reduced one, can be expected.

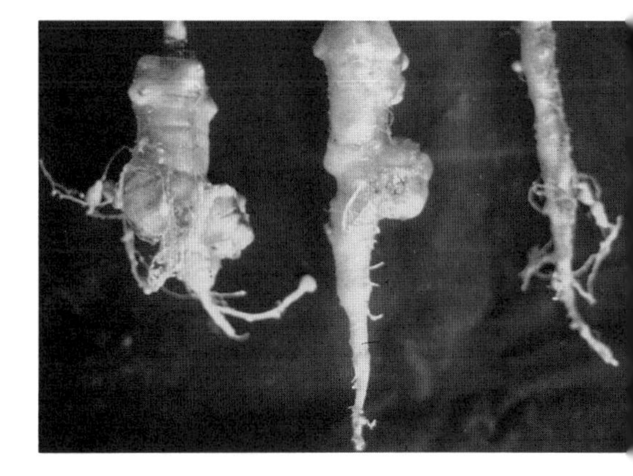

Nematodes are tiny soil organisms which burrow into the roots of susceptible plants. Here the characteristic nodules caused by these pests can be seen in carrot plants. Potatoes and tomatoes are similarly affected.

A white-fronted chat practises some biological control of insect pests and feeds the family at the same time. (PHOTOGRAPH: NORTHSIDE PRODUCTIONS)

Controlling nematodes is quite difficult, but there are some chemical and organic remedies which will produce reasonable results.

### Rodents and birds

Controlling snails and slugs has been dealt with in the last chapter. Rodents and birds are a different matter again.

Most of us are keen to attract birds into our garden, with justification, because they are after all one of nature's most effective pest control methods. They can cause some minor damage, but I consider this a small price to pay for the benefits that our feathered friends bestow on our gardens. The green-fingered gardener will be constantly on the lookout for damage caused by birds and a gentle pruning behind the damage will usually prevent fungal infection.

Rodent damage can often fall into the same category, since some of our native rodents also help to keep down pests in the garden. However the troublesome rodents such as rats and mice may be a problem in some gardens and these should be controlled by using one or other of the rodenticides on the market. A good cat or dog may also help.

## Diseases caused by pathogenic plants

Whilst this may sound like something you would read in a medical encyclopaedia it isn't quite as daunting as it at first seems. Pathogenic plants are simply plants that can cause a disease. Remembering that a disease is defined as any alteration that interferes with the normal structure, functions, or economic value of a plant, pathogenic plants can fall into a number of different categories.

## Parasitic plants

Parasitic plants, such as this mistletoe growing on a eucalypt, can cause damage to host plants. Hand removal is the only effective control method for mistletoe. (PHOTOGRAPH: NORTHSIDE PRODUCTIONS)

Parasitic plants such as dodder and mistletoe can cause severe damage or even death to their host plant.

Dodder is a leafless plant that grows from seeds dropped by birds or carried to the area by animals. The seeds germinate and the stems grow up the host plant. Once established on the plant, the dodder does away with its soil root system and lives entirely off the host plant. Because the dodder is leafless it is unable to manufacture any of its own sugars and starches by means of photosynthesis, so these must also come from the host plant. In order to achieve this, the parasite inserts tiny roots, known as 'haustoria', into the sap stream of the host.

Chemical control of dodder is quite impractical, for a number of reasons. In the first place there are no leaves through which any applied chemical may be absorbed and secondly, since the haustoria have penetrated the sap stream of the host plant, any applied chemical will find its way into that sap stream and could cause the death of the host as well.

The only effective control of dodder is hand removal of the parasite. This can be a laborious task and, in severe infestations, it will often be necessary to remove the host plant as well, since the haustoria will often send out new growth. All the dodder and host plant material should be disposed of so as to avoid reinfestation. In earlier times gardeners burned this type of material, but in today's enlightened society we must find another method of disposal. When the quantity of dodder is not large, place the material in a garbage bag and tie the top tightly. Leave it in the sun for a few days and then send it off to the refuse tip.

Mistletoe seeds are generally deposited by birds on the limb of a host plant. They germinate and then send root-like growths into the host for supplies of food and moisture. Hand removal is the only effective control method and this will possibly involve removal of the affected branch.

Mistletoe is often found near the top of the tree and therefore well out of reach of the average gardener, but in most cases the tree will survive quite well despite the presence of the parasite.

Garden hygiene is an important element in the control of leaf curl, a common disease of stone fruits. Gather up all fallen leaves and either bury them or seal them up in a plastic bag and send them to the tip.

## Lichen

Lichen is another plant that may cause disease, although this is unlikely. Lichens are common on older trees, particularly those that are growing in shaded or damp conditions. I have seen many old trees that have been almost covered by lichens and both the host and the lichens are growing happily.

Improving the light penetration and air circulation will often inhibit the growth of lichen and, if you wish to eliminate it from your garden, a spray or two with one of the copper compounds will usually do the trick.

## Bacteria

Bacteria are tiny, single-celled plants, usually rod shaped, which can cause diseases in other plants. Unlike viruses, these infection agents can live on the outside of a plant rather than inside a cell.

Diseases caused by bacteria produce a variety of symptoms, such as fruit rots, leaf spots, blights, galls and cankers, in a wide range of host plants.

Bacteria may be carried to the host plant by just about any means. Animals, wind, flowing or splashed water, birds, insects, garden tools and equipment or even the gardener may carry the bacteria to another location. They enter the plant through any available opening, such as the stomata on the leaves or any injury site or wound. Rough pruning cuts may provide a possible point of entry for bacteria.

Control is almost as difficult as in the case of viral diseases. Garden hygiene and care when using garden tools and equipment are an important requirement, as is the control of possible insect vectors. It is also a good idea to plant disease-resistant varieties and to rotate your crops.

Chemical sprays are not usually effective in the eradication of a bacterial disease, but spraying with a copper compound can act as a preventative on certain plants.

## Fungi

Fungi come in various shapes, sizes and colours. As with insects, there are both good and bad in the world of fungi. Probably the most common beneficial fungi that we see are mushrooms, which many of us enjoy, but there are others that we find in our everyday existence. The blue and green moulds that we see on decaying food such as bread and the toadstools we often see in garden beds are fungal growths.

Of the many thousands of species of fungi most are harmless saprophytes. In other words, they live on dead or decaying organic matter, and usually benefit the gardener by breaking down material into simple elements which can then be restored to the soil and used by plants. Because of the tremendous importance that these organisms have in the life cycle of the organic matter in the soil, it is vital that gardeners take every precaution to protect saprophytes. Many of these beneficial fungi have been eliminated from gardens by the indiscriminate use of fungicides.

Some fungi are parasitic on nematodes and certain insects which help to control these pests. In people, disagreeable skin diseases such as ringworm and athlete's foot are fungal infections.

Fungi lack chlorophyll and most carry out reproduction by means of spores. They are made up of single threads, called hyphae, and the entire mass of connected hyphae that make up the body of the fungus is called a mycelium.

Not all fungi are a problem in gardens, and some, like these bracket fungi found growing in a Queensland rainforest, are quite beautiful.

Fungi carry out an important role in the decomposition of organic matter. In this instance the fallen log will be broken down over a period of time.

The spores are produced and released from structures known as sporangia. Again, in simple terms, the spores are held in these sporangia until they are mature enough to be released. They are carried from their parent to other parts of the same plant or to other plants, where they settle on the surface or enter the plant tissue through an open wound. The spores are very small, very light and often surprisingly resistant to heat and cold. Because they are able to remain viable for quite a long time, they may be transported over long distances, by wind, insects, animals or human beings.

Once the spore has been deposited on a suitable host plant the fungus must now penetrate the plant tissue in order to obtain its food supply. The spore germinates and a root system similar to that of a plant develops. This root system, or haustoria, enters the plant through an opening such as a wound or through the stomata. Once the fungus has become established and is being provided with a regular and plentiful supply of nutrients and water it will quickly spread throughout the plant and the plant will suffer.

Fungi are capable of producing an enormous number of spores, so the rate of further infection is often dramatically high and fast.

The symptoms of fungal infection are numerous and often very characteristic, so they are easy to identify. Powdery and downy mildews and leaf spots, such as target spot on potatoes and tomatoes, are caused by fungal infection. Fusarium wilt is a fungal disease in tomatoes in which affected plants wilt seriously following infection. This wilting of the leaves could be caused by the physical blockage of the plant's water-conducting tissues by fungal threads.

The powdery mildews affect the surface of the plant's foliage. The leaves are generally covered by a fine white cottony material which makes the leaf look as if it has been dusted with a fine, grey or white, ash-like powder. With downy mildew the inside of the leaf is affected, as well as the surface, on which a soft downy material is evident.

Rusts are another group of fungi and these reveal themselves as a mass of rust-coloured spores which rupture the epidermis or outer skin of the plant, forming small pustules.

Fungal diseases may be controlled in a number of ways. One of the most important is the quarantine regulations which are designed to keep many of these diseases out of a region or country. Eradication of diseased crops or plants, along with the removal of weeds and other aspects of garden hygiene, also play an important role in the control of fungal diseases.

Remember that once a plant leaf is affected by fungi, the damage will always be evident. Fungal diseases are rarely cured, so they must be prevented. Chemical and non-chemical controls usually rely on the fungicide forming a film on the surface of the leaf, preventing any spores deposited there from germinating.

The only fungal diseases that can be cured by chemical or non-chemical sprays are the powdery mildews which, as we have seen, grow on the surface of the leaf and as a result may be killed by an application of fungicide. All other fungi grow within the tissue of the leaf and as fungicides are not penetrative, they do not come in contact with the fungi and are therefore unable to kill it.

Systemic fungicides are available and these are carried around and within the plant by the transportation mechanism. They are therefore placed in a position where they are able to kill the fungus as it penetrates the plant sap stream.

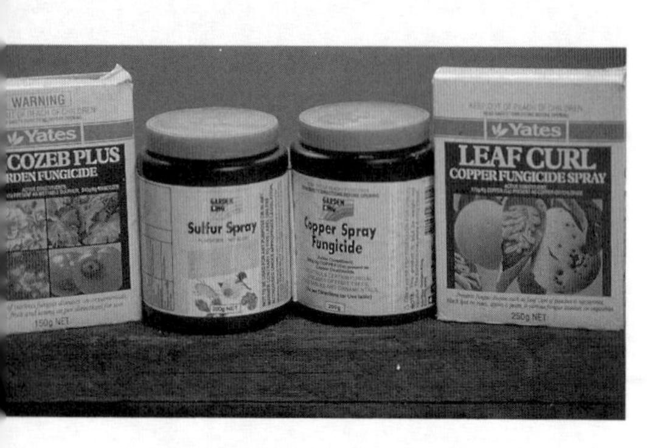


These fungicides are non-systemic, being based on ingredients such as copper, sulphur and mancozeb. They are just a few of the fungicides available to the gardener.

# Some common diseases and their control

## Diseases of foliage, stems, fruit and flowers

### Anthracnose

Anthracnose attacks a wide range of plants but is most common in avocados and beans. The first symptoms on avocados are light brown spots which appear on the fruit. These get larger and darker and the flesh of the fruit generally rots beneath the spot on the skin. Chemical control involves spraying with Mancozeb Plus, zineb or sulphur, while non-chemical control simply involves judicious pruning of diseased parts.

With beans, dark brown marks appear on the stem of young plants that are affected by anthracnose, while on older plants the spots may be evident on leaves, stems and pods. The veins may become black on older leaves. In order to control this disease, it is necessary to spray with Mancozeb Plus or zineb. For those who do not wish to use chemicals, it is essential not to touch or brush up against the beans while the plants are wet so as not to spread the disease, and to save only seed from disease-free plants.

Mangoes are particularly susceptible to anthracnose and the leaves and fruit are both at risk. Dark brown to black spots appear on the leaves and fruit at an early stage and in severe cases fruit will drop some months before maturity. The disease spreads from affected leaves and other parts of the plant to previously unaffected parts. The spores are carried by wind or water. Control is difficult and requires regular and early spraying with Mancozeb Plus. Begin spraying as soon as the flowers form and repeat at fortnightly intervals to within a few weeks of harvest. Those who do not wish to apply chemicals can rely only on pruning and scrupulous hygiene techniques. Gardeners who are still able to burn are advised to rake up all affected leaves and burn them under the tree itself. This destroys the spores on affected leaves, and the smoke wafting through the tree seems to have a preventative effect.

Other plants that may be affected by anthracnose include macadamias, lettuces, poplars, roses and tomatoes.

### Apple scab

Apple scab causes darker green spots to appear on the surface of the fruit. These later turn black and develop corky centres, then the fruit begins to crack. This disease requires regular and well-timed spraying with a copper compound.

As its name suggests, this disease attacks apples, causing spots to appear on the surface of the fruit. These spots are a darker green than the surrounding areas and they gradually turn to black, with corky centres. The fruit then develops cracks. Minor infections are probably not worth worrying about because the fruit is still quite edible.

This disease requires regular and well-timed spraying with one or other of the copper compounds. The first spraying should begin at green tip stage,

Papaws are susceptible to attack by a number of diseases, including dieback, which affect both the leaves and fruit. Because the tree is generally too high to spray, an easy option is to cut it down and cover the top to prevent water from gaining entry to the cut stump. Here a novel method has been used: covering the stump with aluminium foil.

normally early to mid September. A repeat spray should be made in fourteen days time. From then continue spraying every fourteen days with Mancozeb Plus until the end of October. Rake up leaves in autumn and either bury them or place them in large bags and send them to the tip.

## Bacterial diseases

As mentioned earlier in this chapter, bacterial diseases can cause rots, leaf spots, blights, wilts and cankers. Most of these diseases are favoured by high humidity, high temperatures and poor air movement. The major source of infection is other infected plant material and the diseases are spread by dirty garden tools and equipment, insects or contaminated water or soil. Plants that may be affected include mangoes, peas, beans, mulberries, stone fruit, tomatoes, ornamentals, vegetables and tuberous or bulbous plants with a fleshy base.

These diseases are quite difficult to control, but garden hygiene is the most effective method. Preventative spraying with fungicides such as copper will have a marginal effect on some plants. Commercial growers fumigate seeds and grow resistant stocks as a means of minimising the effects.

### Blossom end rot

This disease mostly affects tomatoes and capsicums. The apex of the fruit blackens then splits, and the fruit rots from those splits. The causes of blossom end rot are either a deficiency of calcium, irregular watering, an imbalance between calcium and the other plant nutrients or a combination of all three. Once the disease is evident, it is generally too late to remedy the situation as far as that crop is concerned, so any action that is taken will be for the ensuing crops.

Apply dolomite or lime about six weeks before planting or sowing seeds and water regularly.

### Black spot

Roses are a common target of the fungal disease, black spot, especially in warmer climates. Since surface moisture on the leaves encourages black spot, water roots and soil only, preferably in the early morning, and encourage air circulation around the plant.

Black spot may attack a number of plants, ornamentals or vegetables; in fact any garden plant may be at risk from this disease. Roses are the number one target, however, particularly in the warmer parts of Australia. Queensland rose growers must constantly wage war on black spot. The disease announces its presence by causing an almost circular black spot to appear on the leaves of affected plants. These black spots are often surrounded by a yellow 'halo', which in turn quickly spreads over a large area of the affected leaves, which drop off after a short time.

Black spot is more pronounced during periods of damp, humid weather and surface moisture on the foliage is a vital ingredient for the proliferation of the disease. A lack of air movement will also encourage its formation.

In areas prone to black spot you should spray with a systemic and non-systemic fungicide. Use two consecutive fortnightly sprayings of copper spray, followed two weeks later with a spray of Triforine. Continue this programme right through the spring, summer and autumn months.

As with all other disease-control programmes, garden hygiene is extremely important, and I cannot emphasise this point strongly enough. Regularly, even daily, go around the rose bush and gather up all affected leaves from the ground. Place them in a plastic bag and tie the top tightly. Leave the bag in the hot sun for a few days and then put it in the garbage or tip. Under no circumstances put leaves affected by black spot in the compost heap or bin.

Watering techniques are also important in the control of black spot. To minimise the risk of this disease, never water the foliage in the late afternoon or evening as a surface film of moisture will adhere to the leaves and provide an ideal environment for the black spot fungi to proliferate. Where possible, install a drip watering system around the base of the rose and use this to water the roots and soil only.

Attention to diet is an important aspect of the health of human beings, and the same is true of plants. Attention to the nutritional requirements of the rose will minimise the risk of black spot and other diseases. Most rose fertilisers carry a well-planned programme on the pack, which you should follow exactly.

## Citrus scab

This disease causes a scabby, wart-like growth to appear on the skin of the fruit. Although unsightly in many cases, the fruit is quite OK to eat. Spray with copper and White Oil in the early spring, when about half the petals have fallen from the flowers. A repeat spraying of the fruit in February, with a fungicide such as zineb, will complete the control programme. Always remember that White Oil has the ability to burn if sprayed during the heat of the day in the warmer parts of Australia, so choose the late afternoon or early evening to spray.

## Brown rot

Stone fruit trees are the worst affected by this disease and both flowers and fruit may be damaged. A small brown spot develops and quickly spreads and the flowers or fruit that have been attacked will rot then drop to the ground. To control this disease spray the trees at flowering with either copper or Chlorothalonil. You will need to repeat spray at two-weekly intervals.

Again, hygiene is important, so pick up and destroy all infected flowers and fruit. Pruning to let in light and air will also help.

## Collar rot

Trees affected by this disease may have splitting of the bark at or near ground level. In some cases a sticky gum will ooze from the bark and the bark will lift slightly. There might be a lot of ants around, but they are not the cause of the disease — they are simply using the lifted bark as a home. Foliage on the same side of the tree as the splitting bark may wilt and even die. In some cases the splits may not be near the ground but up on some of the limbs, in which case the foliage on that limb will generally die quite quickly.

The disease is encouraged by damp soils and a lack of air or light. It can be controlled in the early stages by taking a sharp knife or chisel and trimming back all damaged bark to good wood. Mix up a paste of copper fungicide and paint this paste over the affected parts. Make sure that you brush the paste firmly into any crevices in the surrounding bark so as to guard against a reinfestation. If more than 60 per cent of the circumference of the trunk or limb is affected, then the removal of the bark will effectively ringbark the tree. Obviously when this occurs, the tree is going to die. The only remedy is to remove the tree in this case.

## Leaf blights

Some potatoes, celery, carrots, strawberries, corn and a range of ornamentals are susceptible to leaf blights. These diseases cause spotting or discolouration of the leaves, with black patches or spots appearing within the discoloured areas. Leaf blights are usually more prevalent in damp conditions and can

### Condys crystals

These are useful, both as a fungicide and as a means of controlling worms in pot plants. Dissolve at the rate of one gram per litre of water and spray over plants affected by powdery mildew. Pour this same mixture through potting soil and worms will rise to the surface where they can be picked off.

### Chamomile tea

Chamomile has been referred to as 'the garden doctor' because of its beneficial effects on a number of plants and its ability to control and inhibit various fungal diseases. Damping off in seedlings is one problem that can be controlled by chamomile tea, as are the downy and powdery mildews. Simply cover a handful of chamomile leaves with boiling water, allow to cool, strain off the liquid and apply it to the plants concerned.

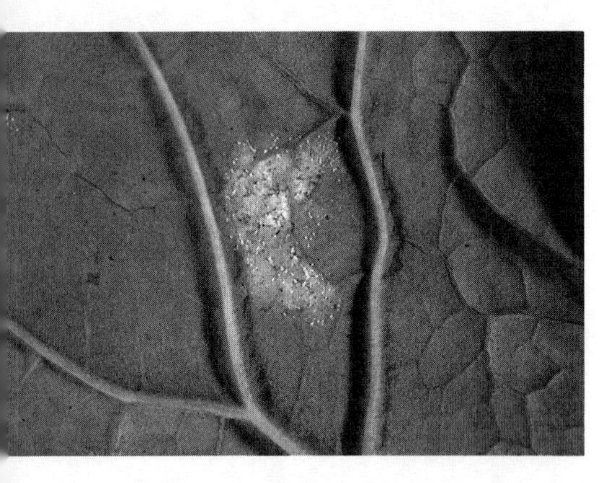

A patch of downy mildew on a broccoli leaf. Damp, humid conditions favour the spread of this disease.

seriously disfigure ornamentals or reduce yields of fruit and vegetable crops. Control by spraying with copper, Mancozeb Plus or zineb. Repeat sprays may be necessary.

## Leaf curl

This well-known disease is prevalent in all stone fruit, but mostly in peaches and nectarines. The leaves thicken and curl and blister-like growths appear. Sometimes these distortions are pink, but they may also be white or even green. After some time the leaves fall off and die. Fruit may also be affected by this disease.

Infection occurs at the time of bud burst — the time when the buds swell to their limit before bursting into leaf. Begin control spraying in late winter and again at, or just before, the buds start to swell. Spraying with copper compounds is the usual method of control.

Once again garden hygiene is an important element in the control or reduction of leaf curl. Gather up all fallen leaves and either bury them or place them in a plastic bag, tie it tightly and send it to the tip.

## Leaf spots

Spots of varying shapes and sizes, mostly brown, appear on the foliage of affected plants. Some of these spots may have a dark brown ring as a halo around the edge of the spot. As the disease progresses, these spots may join together and large portions of the foliage can be affected. Growth and yields are both influenced because of the reduction in the leaf's ability to produce sugars and starches through photosynthesis. Peas, bananas, berry fruits, ornamentals, perennials and fruit trees may all be affected by leaf spots.

Begin spraying with copper, Mancozeb Plus or zineb as soon as the first spot appears.

## Mildews

Powdery mildew starts off as a faint greyish or white spot on the upper surface of the leaf, then a grey, ash-like, powdery growth forms and eventually covers the upper surface. You rarely find any sign of damage on the underside of the leaf until the disease is well advanced. Dry and warm weather favours the spread of this disease, so it is more prevalent during the drier months.

A wide range of plants are affected, but peas and beans, Crepe Myrtle, papaws, roses, apples, grapes, lettuce, cucurbits, strawberries, many ornamentals and flowering annuals are all particularly vulnerable.

Sulphur sprays are effective in the control of powdery mildews, but take care not to apply them during the warmer weather. Rockmelons, watermelons and other members of the cucurbit family are susceptible to sulphur damage, so it is wise to avoid spraying these plants with that compound. Triforine sprays are effective in the control of powdery mildew but there are few that are registered for this use in gardens. Mancozeb Plus is another that may be used to control this disease.

Downy mildew appears on the upper surface first, but most of the furry growth appears underneath the spots that can be seen on the top of the leaf. Damp, humid conditions favour the spread of this disease. The affected spots turn brown or reddish brown and the foliage dies quite quickly. Vegetables, ornamentals and fruit are all susceptible to attack from downy mildew. Spraying with fluraxyl or zineb will control this disease.

The symptoms of downy mildew, to which grapes are particularly prone, first appear on the upper surface of the leaf, but the furry growth appears underneath the spots.

### Mosaics

These are viral diseases which attack ornamental plants and vegetables. Tomatoes, potatoes, cucurbits, cabbages, cauliflowers, roses and camellias are prone to attack. These viral diseases can display different symptoms, but the most common is a mottling of the leaves between the veins.

Controlling viral diseases has been mentioned earlier but it is worth repeating. Make sure that virus-free plants are used for propagation and control insect vectors such as aphids, thrips and mealy bugs by spraying with an appropriate insecticide.

Rust appears as a yellow spot or patch on the upper surface of the leaf, as can be seen on these pelargonium leaves. Directly below these spots, on the underside of the leaf, rusty brown pustules develop. (PHOTOGRAPH: ALLEN GILBERT)

### Petal blight

This is a particularly troublesome disease of azaleas, especially the evergreen Kurumes and Indicas that are growing in warmer climates. The flowers turn brown and papery shortly after opening and stay in that dead-looking condition on the plant for some weeks. This gives the azalea bush an extremely unsightly appearance and spoils the effect that these wonderful plants can provide.

Control is very difficult and you can only achieve reasonable success by using chemicals. As soon as the buds start to show some colour, spray with triadimefon. Repeat sprays will be necessary. It has been suggested that spraying the bush and the surrounding soil with a copper spray in January may also help. In New Zealand Bravo is registered for the control of this disease.

### Rust

There are many different fungi that may cause the disease called 'rust', but the end result is always the same. A yellow patch or spot will appear on the upper surface of the leaf and on close inspection, you will notice rusty brown pustules on the underside of the leaf directly below the yellow spots. The rust spores are usually blown around by the wind. This disease affects a wide range of ornamental and fruiting plants, vegetables and perennials. Some of the more common examples of rust can be seen on poplars, willows, gerberas, beans and marigolds, to name just a few.

Spraying with Mancozeb Plus, zineb or sulphur sprays are the common chemical controls for rust; while pruning and destroying diseased foliage, along with other recognised garden hygiene techniques, are the non-chemical methods of control.

## Lawn diseases

### Brown patch

This disease appears during the late summer and autumn when the weather is hot and humid. Large brownish patches, almost circular, appear in various parts of the lawn. Unfortunately this disease usually appears at the same time that lawn grubs are at their most active and gardeners have difficulty in identifying the culprit. In order to decide whether it is brown patch or lawn grubs, lay a wet bag or piece of carpet on the lawn overnight, as described in the previous chapter. When it is lifted in the morning the lawn grubs will be on the surface of the grass. If there are no lawn grubs in sight, it's a fairly safe bet that the damage is being caused by brown patch. Spray with chlorothalinol or Mancozeb Plus.

### Dollar spot

Areas of lawn shaded from the early morning sun are most at risk from this disease which generally occurs in late winter or early spring. It takes its name

**Lawn tip**

**Disease prevention in lawns will be enhanced by attending to some basic mowing principles. Close mowing removes leaf area that would otherwise produce sugars and starches through the process of photosynthesis. By raising the cutting height of the mower a larger leaf surface will be exposed to the light, resulting in a greater supply of essential food, which in turn means a healthier, more disease-resistant lawn.**

Whilst the *Hardenbergia violacae* (Purple Coral Pea or Happy Wanderer) enjoys regular watering during dry periods, it does not enjoy consistent heavy rain which, in this instance, has brought on an attack of phytophthora or root disease.

from the fact that it begins as a small, almost circular, brown patch which American gardeners thought resembled the shape and size of their dollar coin. It is often possible in the early morning dew to see the fine, white, cobweb-like threads of the fungal mycelium just below the tips of the grass. If you can see them, brush the lawn with a broom to slow down the rate of damage that the fungus will cause, although your neighbours are likely to be completely entranced by the fact that you are out there sweeping your lawn! It will be necessary, however, to follow this up with a spray programme using Chlorothalinol or Mancozeb Plus.

In both cases, an excessively high level of nitrogen in relation to the potassium levels will increase the incidence of disease. Potassium encourages disease resistance and so it is important when fertilising a lawn to ensure that adequate potassium levels are maintained. Lawn greeners that are composed entirely of nitrogen are not, in my opinion, the most suitable for this purpose. I would strongly recommend that gardeners maintain vigorous lawn growth throughout the growing period by using a fertiliser that contains both nitrogen and potassium, such as Supergreen.

## Soil-borne diseases

### Phytophthora

This disease attacks the roots of plants and is encouraged by damp soil conditions. All plants, whether they are in soil or pots, are at risk and the disease results in death, more often than not. In its early stages, phytophthora causes the foliage to wilt. Early leaf drop may occur in some cases, but in most, wilted or dead leaves hang limply on the tree or bush for some time. In the case of minor infections, growth may be severely inhibited and the plant may have sparse foliage.

Control involves a range of activities. First, improve the drainage and avoid over-watering, as both are common causes of the problem. Keep mulches away from tree trunks and avoid damaging the roots of shallow-rooted plants by cultivation. A drench with fluraxyl will also help control this disease. Repeat drenching may well be necessary and is essential if a new plant is to be put in the same spot as one which has died from this disease.

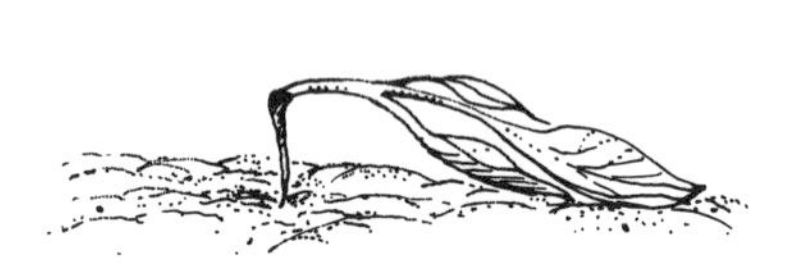

pythium causes seedlings to rot off at ground level

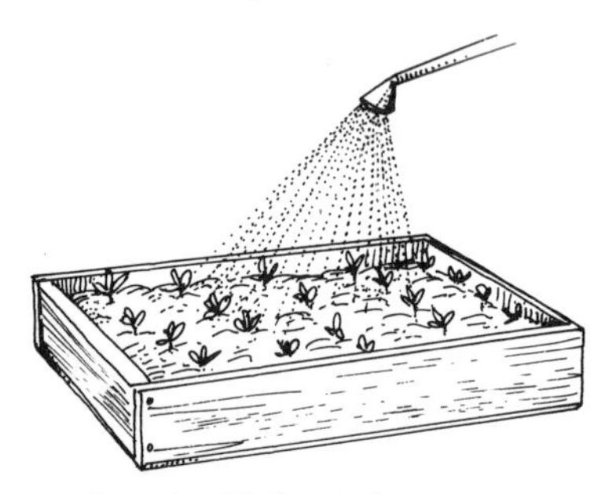

drench with fluraxyl as soon as seedlings emerge

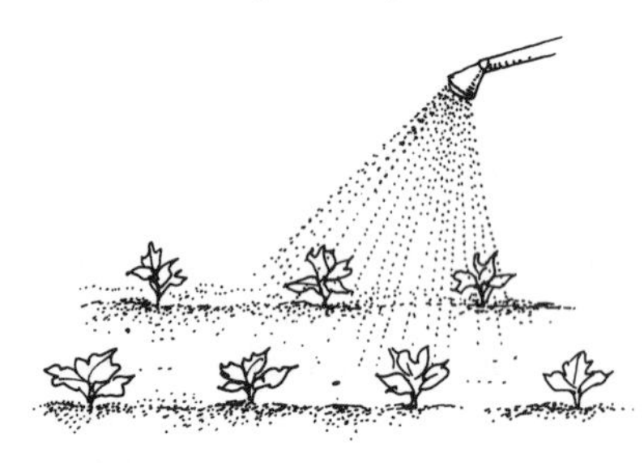

drench again when transplanting

Pythium is a disease of very young plants which causes seedlings to rot off at or just above ground level, sometimes overnight. To prevent it, drench the potting mix in which seedlings are growing with fluraxyl as soon as the plants emerge, after pricking out and at the time of transplanting.

### Pythium

Seedling growers need to be on the lookout for this disease of very young plants. It causes the seedlings to rot off just at or above ground level. It can happen overnight and is often confused with the damage caused by the cut worm. You can tell the difference, however, because you can see that the stem has rotted through if pythium is the problem.

As a preventative measure, drench the potting mix in which the seedlings are being grown with fluraxyl as soon as possible after emergence or pricking out, and drench at the time of transplanting.

### Tomato wilt

This fungal disease enters the tomato plant through the roots and affects the movement of sap and nutrients via the transportation system. The tomato plant will start to wilt on hot days, due to the lack of turgor pressure (the cellular pressure in the plant), and the lower leaves will turn yellow from lack of nutrition. Before long the whole plant dies and needs to be pulled out of the ground. To identify this disease accurately, split the stem of affected plants vertically and look for a brown discolouration beneath the green outer skin of the stem.

This degree of leaf damage can be caused by tissue-feeding insects in a fairly short time. (PHOTOGRAPH: NORTHSIDE PRODUCTIONS)

Most organic gardeners encourage lizards, since they play an important role in the control of many of the smaller insects. (PHOTOGRAPH: NORTHSIDE PRODUCTIONS)

Some insects form a sticky web with which to glue together foliage, thereby providing protection from birds and other predators. The sticky web may be seen on the leaf in the centre of the photograph.

Many plants are unable to cope with warm summer rain when they are grown outside their normal climate. This *Leptospermum scoparium* (Manuka) has suffered badly as a result of a wet, humid summer.

The devastating effect of grey mould, or *Botrytis cinerea*, on a rose. Since the mould very often starts on dead tissue such as spent flowers, this is another good reason for deadheading your roses. (PHOTOGRAPH: ALLEN GILBERT)

Aphids can generally be found near buds or the growing points of plants. In this case they are congregated on the flower stem, where sap movement is at a maximum. This is also a nice soft spot which the insect can pierce with its proboscis.

As the spores can last in the soil for some years, this is a very difficult disease to control and crop rotation is essential.

## Disease prevention

It has been said that there are many similarities between plant diseases and the diseases of humans, both in their causes and effects. While we, as gardeners, cannot be expected to understand the intricacies of human disease, we can and should understand the basic principles of plant disease, prevention and cure.

Prevention of plant disease is an important element of garden management. The reckless and indiscriminate use of plant fungicides is neither good for the garden nor the gardener. It must be remembered that once a plant leaf shows signs of attack by a fungal disease, those signs will stay on that leaf until it drops off the plant (the only exception to this rule is powdery mildew). So it is the wise gardener who sees to it that disease is prevented before it defaces his plant.

As has already been mentioned, the first element of disease control is to maintain an adequate level of nutrition. I hold firmly to the belief that it is far better to spend an extra dollar on fertiliser than it is to spend an unnecessary dollar on fungicide. I emphasise the word 'unnecessary', as it is impractical and illogical to suggest that fungicides are *never* going to be needed. The point I'm making is that, with adequate nutrition, they won't be needed as much.

Adequate nutrition and watering techniques have already been mentioned and these should be carefully attended to in order to minimise the incidence of fungal disease. As we have seen, you should not water in the evening, especially in summer, since free water sitting for long periods on foliage during the warm summer nights provides an ideal environment for the germination of fungus spores. From a disease point of view, then, it's best to water plants very early in the morning. Unfortunately this is not the best watering period from an evaporation and water loss point of view. So, mulching will help conserve water applied in the early morning. If watering in the evening is unavoidable, ensure that only the roots are watered and not the foliage.

Certain weeds provide the habitat necessary for certain stages in the life cycle of the fungi, so you need to keep your garden weed free to eliminate fungal diseases. This aspect of garden hygiene is as important for fungus control as it is for insect control.

Despite all these precautions, diseases will still occur so it is important to know how to counteract them. In the majority of instances you will need to use one of the fungicides that are available.

### Fungicides

Many home gardeners are indiscriminate in their use of home garden chemicals and this can sometimes result in the product not doing the job it was designed to do. As mentioned earlier, fungicides act mainly as a preventative and you should use them accordingly. This means that you should apply them just before the period when fungal attacks are known to occur and maintain them until the likelihood of attack has passed.

Continual use of the same fungicide has caused certain fungi to build up resistance. This is dealt with later in this book, but the use of alternate fungicides in a spraying programme will go a long way towards minimising this risk. Use fungicides wisely and in accordance with the directions and plant diseases should not present too much of a problem in the garden.

One final tip: dip your secateurs in a solution of disinfectant or straight methylated spirits between each pruning cut. By doing this, the chances of transferring diseases from one part of the plant to another will be minimised.

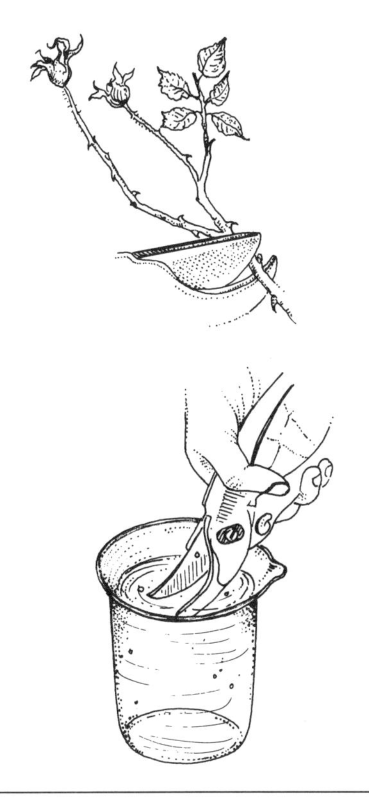

Dip your secateurs in a disinfectant solution or methylated spirits between each pruning cut to minimise the spread of disease from one part of the plant to another.

# Using chemicals wisely

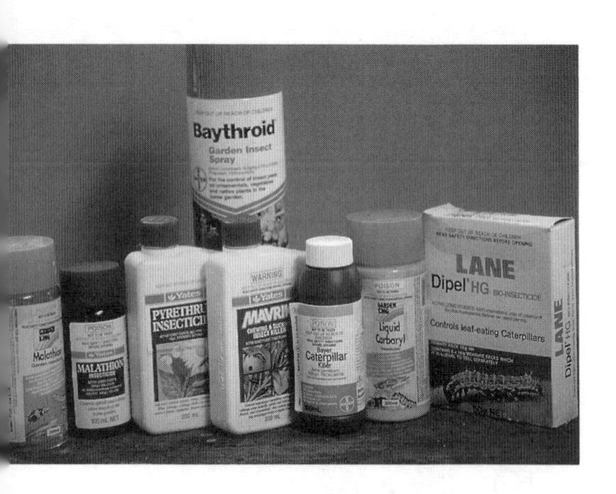


This is only a small portion of the range of insecticides available to the home gardener. Spend some time reading and absorbing the information provided on these packs. It will help you to make a responsible decision about which one to use.

## Poisons

The toxicity of chemicals varies considerably and is described by the 'LD50' measurement. When a certain dosage of an active ingredient is given to a group of test animals, the quantity required to kill 50 per cent of them is called the LD50 measurement. The higher this value, the lower the toxicity of the chemical.

The Poisons Schedule is different from the LD50 measurement, even though it is partly based on those values. The schedule associated with gardening products covers Schedule 5, which is slightly toxic to humans; Schedule 6, which is moderately toxic to humans, and Schedule 7, which is extremely toxic to humans. Regardless of the schedule number, though, gardeners should always treat chemical products with respect and take the necessary precautions.

Pesticides may enter the body by one or all of the following methods:

- Skin contact — called 'dermal', the chemical is absorbed through the skin.
- Inhalation — breathing in fumes, vapours, or dusts.
- Oral — eating, drinking or swallowing.

### Dermal

Most poisonings occur as a result of dermal contact, either through spillage of chemicals over certain parts of the body, or in the case of vine crops, by workers brushing against sprayed plants. Of course this could happen with any gardener who brushes a bare arm against a plant that has already been sprayed. Spillage often happens when a gardener is mixing chemicals, too, and splashes on trousers during mixing can cause dermal contact with the chemical. The crotch area is very absorbent — more so than the forearm. The face, scalp and abdomen are all very absorbent areas of the body too, and should be protected at all times. The lesson here is *always* to wear protective clothing.

### Inhalation

Inhalation is more common with dusts and powders, but can certainly occur with vapours and fine mists. Mixing chemicals indoors increases your chances of inhaling dangerous vapours or dusts, especially if you do it without a respirator or mask. Always mix or spray chemicals with your back to the breeze, and try not to tackle any spray job on very windy days. *Always* wear a mask when mixing and spraying.

### Oral

It is usually children under five years of age who swallow insecticides or fungicides accidentally and this is the result of neglect or ineffective storage. I will discuss safe ways to store garden chemicals later in this chapter.

## How to protect yourself

### Read and understand the label

It is not enough to cast a cursory glance over the writing on the chemical pack. Admittedly the print is usually fine and there seems to be so much of it, but remember that it is put there for your protection. Don't just read the amount of chemical that must be mixed with a given quantity of water, read the safety precautions and the warnings as well.

State and federal government regulations control the labelling of agricultural and garden chemicals strictly. Labels must include such information as the concentration of chemical to be used, the plants which it is safe to use the chemical on and the precautions that should be taken to avoid harm to plants or animals. If the product is a poison, it will have the schedule number printed on the pack. First aid advice is also listed — make sure you read it.

### Wear protective clothing

When mixing or spraying garden chemicals, it is essential to wear protective clothing. Gloves are mandatory and they should be rubber, not cotton or leather, as these will absorb chemicals. For the same reason gloves should not be lined, as this too will absorb chemical. They should extend well up the forearm so as to prevent the possibility of chemical spilling inside the glove.

Thongs might be ideal for the beach but they are certainly not appropriate for spraying the garden. Gumboots are the best footwear as they will not absorb chemicals and will protect the feet and legs. Wear your trousers or overalls outside the gumboots so that, if you do spill a chemical, it will run down the outside of the boot and not the inside.

As we have already seen, the face is an absorbent part of the body and should be protected as much as possible. The eyes and the nose are particularly vulnerable and should be protected by goggles and a mask. However, these do not protect the whole face and, as you should wear a hat too, consider buying a helmet with a face mask. You might like to wear short-sleeved shirts and shorts for working in the garden, but they are not suitable when you are using chemicals. Wear the relatively cheap overalls which cover the legs and arms and wash them immediately after use. You can also buy disposable overalls which you can roll up and put in the garbage.

### Mix chemicals outside

It is always tempting to mix chemicals in the shed where they are stored, but you need good air movement to prevent you from inhaling vapour. So, always mix chemicals outside, with the container on your down-wind side.

Only mix the amount of chemical you need to do the job in hand. Don't mix up a larger quantity and store it in another container or, even worse, leave the unused portion in the sprayer.

It is essential to wear protective clothing — including long, unlined gloves, a mask or goggles and a hat — when mixing or spraying garden chemicals.

### Apply carefully

Avoid spraying on a windy day as this can cause spray to drift onto you when you get down wind of the target. Even if there is only a slight breeze it will still

Before spraying an insecticide or fungicide, always make sure that there are no traces of weedkiller left in the sprayer. The results of not doing this can be seen in this garden, where valuable roses were sprayed for black spot with a sprayer still containing vestiges of Roundup. It is advisable to have two sprayers — one for weedkillers and one for insecticides and fungicides.

be enough to cause fumes or vapours to drift onto you. Wherever possible, work where the vapours and fumes waft away from you.

### Maintain your equipment

By keeping your spraying equipment in tip-top condition, you will minimise the risk of coming into contact with chemicals. Joints should be kept tight and washers replaced as soon as they start to lose their sealing ability. Many of the sprayer wands that I have seen are in desperate need of attention as they leak chemicals over the hands while under pressure. Replace hoses regularly to minimise the risk of one bursting and spraying you with an insecticide or fungicide mixture.

### Keep children and pets away

Both children and pets tend to be rather inquisitive when it comes to gardening activities and it is always nice to have them around while you're in the garden. However, both should be kept well away from the spraying operation, as they are likely to be affected by the chemicals being used.

### Decontamination

As soon as spraying is finished, the wise and careful gardener will immediately change out of the protective clothing. Items that need washing should be washed straight away, while disposable overalls should be disposed of carefully. Wash respirators, gloves, boots and so on and store them, ready for next time.

Wash out the spraying equipment thoroughly straight after use and put the chemical away in a safe place.

Don't eat, or smoke if you are a smoker, until you have had a shower.

### Last resort

If you should be unfortunate or silly enough to accidentally swallow or come in contact with a garden chemical, read the label directions and warning and seek medical advice in accordance with those directions. Take a note of the name of the chemical before calling or visiting a doctor.

## Withholding period

All garden chemical packs must by law carry a statement advising users of the withholding period of that particular product, but some gardeners do not fully understand the implications of this advice.

The withholding statement will say something like: 'Do not apply later than seven days before harvest' (obviously, the number of days will vary depending on the chemical concerned).

I remember a gardener calling me some years ago to tell me that he had sprayed his tomatoes with an insecticide that had a withholding period of fourteen days. He wanted to know if it would be all right to eat the fruit if he kept it in the cupboard for fourteen days. I told him that this would mean that he would be eating the chemical as well, because the withholding period clearly stated that spraying should take place no less than fourteen days before *harvest*, not before consumption.

Garden chemicals break down over varying periods of time. The agents for breaking down chemicals on the surface of the plant include direct sunlight, general light, rain, temperature and the general decomposition of the chemical itself. Systemic chemicals, and penetrative chemicals to a lesser degree, are broken down by those agents already mentioned, but the concentration of sap must also be considered. When the chemical is applied to the plant, it is taken into the sap stream. As time goes on, the plant continues to produce sap but the amount of chemical stays the same, so the concentration of that chemical becomes smaller and smaller. The time that this dilution takes varies from one day to around seventy-two days, but the gardener needs to know and understand these principles. Wherever possible, choose the chemical with the shortest withholding period. The organic insecticides, such as Derris Dust and pyrethrum, have a withholding period of one day, whereas diazinon has a withholding period of fourteen days.

## Using the right quantity

It is not only wasteful to use more than the recommended rate, it is also harmful to the environment. The rates recommended on the pack are those that have been shown to be effective in trial work done before the product was released. Some gardeners think that if the recommended rate is going to do the job, a little extra will do it even better. This is a classic case of more not being better.

If a chemical containing solvents is used in excessive amounts it can cause phytotoxic reactions in certain plants. In other instances, excessive amounts of the chemical itself may cause damage to plants — often more damage than the pest or disease it was supposed to control.

Gardeners who use less than the recommended rate, on the other hand, are likely to be wasting their money as well, because a too low concentration of the chemical won't work.

## Storage of chemicals

All pesticides should be stored safely and responsibly. It is imperative that all home garden chemicals should remain in their original containers, and that the manufacturer's label should be readable at all times. Some years ago I visited a home gardener who had problems with some trees in his yard. Between the time of his original call and my visit, not only had his trees died, but a swathe had been carved through two of his neighbours' gardens as well. I asked him whether he had carried out any gardening activities that might have caused this problem, but he assured me that the only thing he had done was to spray some weeds around the back of his swimming pool.

At the back of the swimming pool, I discovered that this gardener had certainly killed every vestige of weeds and grass nearby. The pool happened to be at the top of a slope above his neighbour's garden and was the beginning of the avalanche of death to the plants below. He had been given a plastic bag containing a most effective weedkiller used in agricultural and industrial situations. This particular weedkiller can travel through the soil and that's what had happened.

I was most indignant and rather rude to this individual because of his stupidity in using a chemical for which he did not have any instructions. If he had had the original container, he would have been able to read the likely result of its use. I trust that he has learned a valuable lesson from an extremely expensive exercise.

Another common failing with home gardeners is to mix up more of a chemical than they are going to need for the current spraying task. They then put the unused quantity in a soft drink bottle for future use. This is a stupid and dangerous practice. It is stupid because the majority of chemicals break down in a relatively short time when mixed with water, and when the time comes to

This graffitti holds a message for gardeners: treat garden chemicals with great caution and don't push your luck by using, storing or disposing of them carelessly. (PHOTOGRAPH: ALLEN GILBERT)

use them, the chances are they will no longer be effective. It is very dangerous because the chemicals are stored in containers with which young children are often quite familiar. Many accidental poisonings are caused by the irresponsible practice of storing chemicals in soft drink bottles. Young children think that they still contain the soft drink, and adults, too, have been known to make this mistake.

The old tin garden shed is probably the most inefficient building possible for storing home garden chemicals, because it is extremely hot in summer and cold in winter, and temperature variations are not conducive to long life in chemicals. Always store home garden chemicals in a locked cupboard.

Common household chemicals, such as chlorines, bleaches, kerosene, oil, methylated spirits and paints are often stored in the same shed as the garden chemicals and this, too, is dangerous. Flammable materials should also be stored well away from garden products.

## Buying agricultural packs

There is a great temptation amongst gardeners, particularly those in or near the country, to buy the larger agricultural containers of chemicals. These chemicals are more concentrated and go a long way further.

The economics look good and when you consider how far that large pack will go, they look even better. A 4-litre container costs nowhere near as much as eight 500-millilitre bottles of the same home garden product, but believe me, this is false economy.

A 4-litre container of chemicals is usually enough to last the average home gardener about twenty or so years, and there are no chemicals that I'm aware of that can be stored and used for that long and still remain effective. The other problem is that the directions are written with broad-scale agricultural application in mind. I know many gardeners who have difficulty in understanding and adhering to the simple directions on home garden chemical packs, so how on earth are they going to interpret directions that tell them to use so many hundred millilitres in 1000 litres of water and apply over so many hectares of ground?

Sometimes groups of gardeners, particularly members of a specialist society or club, buy an agricultural pack of a chemical and then divide it up between them. This may seem a good idea, and one might assume that as specialists they know how to use garden chemicals, but I have seen many specialist growers making the same mistakes as amateurs. There is also the problem of storing the chemical when your share is transferred from the original container — perhaps in a soft drink bottle or some other container that has no relevance to the chemical it contains. Only one of the group will have the original container with the manufacturer's directions. The others, at best, will copy them and may or may not have them at hand when they mix up the chemicals. This is an unwise practice and should not be encouraged.

Agricultural chemicals are often more concentrated than their domestic equivalents which, when used on a specific range of plants, cause them no harm. The domestic equivalent of a particular agricultural chemical is often aimed at a different group of plants and, as a result, may have a much lower concentration of the same active ingredient. When the agricultural concentration is applied to the domestic plants, there may be a phytotoxic reaction which can cause defoliation or even death. In such a case I would find it hard not to tell the owners of the plants that it served them right for using an agricultural product in a domestic situation.

There is still another disadvantage of buying in bulk. Because the gardener has a large quantity of pesticide, it means that he or she will continue to use

**Handy hint**

When using an aerosol insecticide on house plants always ensure that you hold the can a reasonable distance from the plant, otherwise the propellent gas may cause serious damage to the plant. It is far better to let the insecticide kill the insect pest than to hold the can close to the plant, thereby causing the insect to die of starvation because of the demise of the host plant.

the same product over and over again, so increasing the chances of insect pests and diseases building up a resistance to that particular active ingredient.

It is easy to see that I am totally opposed to home gardeners buying agricultural pack sizes of insecticide and fungicide. They have purchased enough insecticide or fungicide to last them for longer than the life of the product. This means that they will be blissfully spraying a chemical around which has no effect whatsoever on the insects or fungi. If that is anything other than false economy, I'm willing to start walking backwards to Burke any day you like!

Recommended protective clothing. The commercially available overalls pictured are disposable and can be rolled up and put in the garbage after use. (PHOTOGRAPH: ALLEN GILBERT)

## Resistance factor

Resistance to certain active ingredients has become a major problem in pest and disease control and this has usually been the fault of the gardener who over-uses chemicals.

Although I have no intimate knowledge of medicine, I believe that when we go to the doctor for a flu injection, what we get is a mild dose of influenza. This mild dose encourages our immune system to build up antibodies that will fight the infection so that when the real flu attacks, our immune system is on full alert and is able to ensure that we do not succumb to the disease.

The same thing happens when we indiscriminately apply insecticides and fungicides to plants. Imagine this. We spray a systemic insecticide onto our plants for the control of aphids. At the time of spraying the insects were not around but the chemical was absorbed into the sap stream of the plant. With the passage of time, the plant produces more sap and, as we have seen, the concentration of the insecticide is reduced as a result. About this time a swarm of aphids arrive on that plant and set about sucking the sap. Because the concentration of insecticide is now too weak to have any serious effect, the aphids continue to feed and do not die. They have then been exposed to a weak concentration of that insecticide — just enough to encourage their immune system to build up a resistance. Their offspring are likely to inherit that same resistance factor.

So, how does one avoid enhancing the resistance factor of pests and diseases? The answer involves the gardener in a few logical steps.

- Alternate between insecticides when controlling pests and fungicides when controlling fungi. Don't use the same product time after time, as this encourages resistance. By spraying an alternative product, the pests or fungi resistant to the previously-used product will be killed off and the resistance factor broken.
- Spray only when necessary and always use the appropriate insecticide or fungicide for the pest or disease that is being controlled.
- Protect the predators, so they too may kill off resistant insects or fungi.
- Avoid using multi-purpose sprays when a single purpose product will do the job.

This protective helmet for gardeners includes a battery-operated fresh air supply.

## Disposal of chemicals

Empty chemical containers need to be disposed of properly. Rinse the empty container several times with clear water. Pressure cans should be completely empty before disposal and they should never be punctured or thrown into a fire, as this can cause serious explosions. It is best to consign these containers to the tip in your normal household rubbish.

Any unused chemicals should be very carefully disposed of. They should not be thrown into the rubbish with the household garbage, as they could end up as landfill and cause problems. Most local governments have special facilities for the disposal of out-of-date and unregistered garden chemicals.

# Some alternatives

**M**y father was a very good gardener and it was from him that I learned most of my basic gardening skills. His vegetable garden was admired by all who came to visit, and one of the things he practised with great diligence was crop rotation. This meant that he grew different crops in different parts of the garden each year. It would have been unthinkable for him to grow the same crop in the same patch year after year.

## Crop rotation

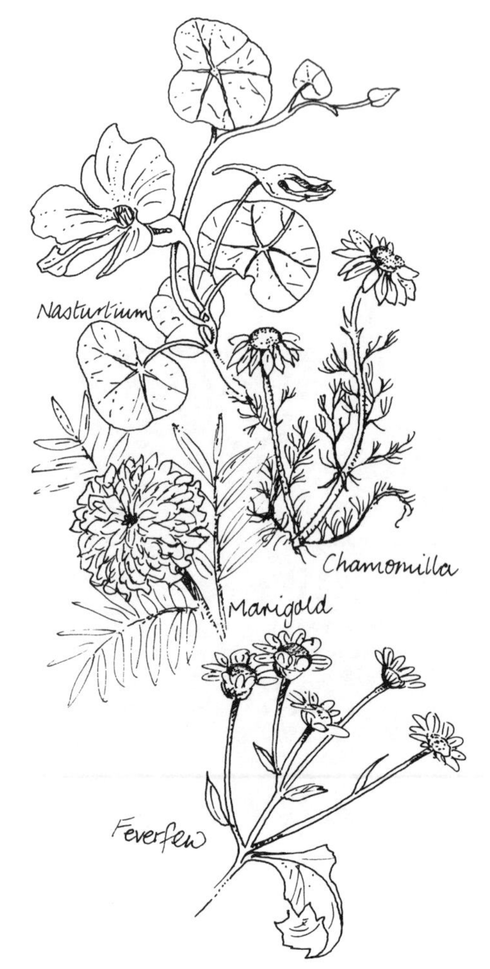


While crop rotation will not eliminate the need for pest and disease control, it does certainly reduce the gardener's dependence on the use of pesticides and fungicides. Certain pests and diseases are specific to a particular plant, but when these host plants are grown in a different area of the garden each year, they find it a lot harder to survive. Other pests and diseases often overwinter in plant debris. If you grow the same plants in the same area year in and year out, there is always enough debris remaining in the garden bed to ensure continued survival of the insect or fungus.

Plant problems associated with pests of limited mobility, such as nematodes, are significantly reduced when susceptible plants are not grown in the affected area for a few years.

Because different groups of plants have different nutritional requirements, the gardener who practises crop rotation ensures that the soil is not being depleted of any one nutrient or group of nutrients. This means that diseases brought about by inadequate soil nutrition can often be avoided.

Working out a crop rotation system is not quite as simple as it may seem, because in the first instance different vegetables require different amounts of room. In order to understand the principles it is first necessary to learn of the plants that are related to each other.

### Related plants

1 Tomato, potato, capsicum, eggplant.
2 The legumes — peas and all the beans.
3 Cabbage, cauliflower, broccoli, brussels sprout, turnip, swede, kohlrabi, radish.
4 Onions, garlic, shallots, leeks, chives.
5 Beetroot, silver beet, turnip.
6 Lettuce, endive, artichoke.
7 Carrot, parsnip, parsley, celery.
8 Pumpkin, zucchini, cucumber, melon, squash, marrow.

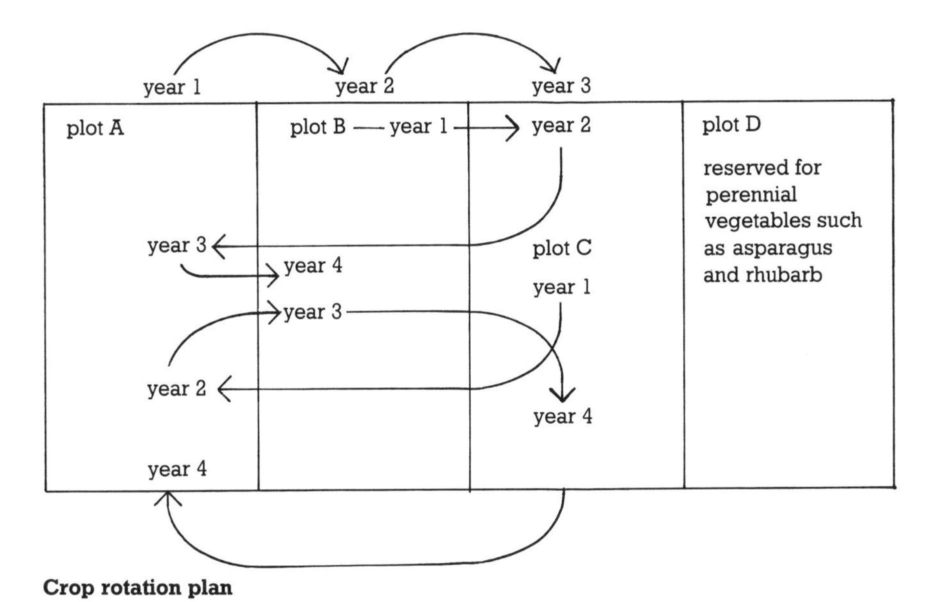


**Crop rotation plan**

A good crop rotation programme will ensure that successive crops of one family are not grown in the same bed for at least a couple of years. In order to maintain a workable crop rotation system, the vegetable garden should be divided up into four plots.

In Year 1 the following vegetables can be grown in Plot A: potatoes, carrots, beetroot, parsnips, onions, leeks, shallots, tomatoes, capsicum, zucchini, marrow, pumpkins, melons, cucumbers.

In Plot B: peas, beans, sweet corn, silver beet, spinach, lettuce, endive.

In Plot C: cabbage, cauliflower, brussels sprout, kale, swede, turnips, radish, kohlrabi.

Plot D is reserved for the perennial vegetables such as asparagus, rhubarb, and so on.

In the second year those vegetables move to the next plot, in other words, those in Plot A go into Plot B and those in Plot B go on to Plot C, whilst those that were in Plot C move back to Plot A. In the third year they all move onto the next plot, which means that it will be four years before the same vegetables are back growing in their original plots.

## Companion planting

Pest and disease control through the use of companion planting is also practised by a number of gardeners, often with less tangible effects than other forms of control. I know gardeners who swear that companion planting has significantly reduced the incidence of a particular pest or disease in their garden, but there are others who will say that it just doesn't work at all.

Companion planting relies on the repellent effects of herbs and other plants on certain insects. Marigolds, for example, are said to repel nematodes by giving off a repellent chemical from their roots. In other cases it is the fragrance of the herb or plant which repels pests. A well-known overseas herb grower was recently quite insistent that companion planting was nowhere near as effective as we have been led to believe, however, and there does not seem to be much in the way of hard and fast scientific evidence to support the practice.

I am sure that there is a direct correlation between the success of the system and the pressure of insects that are present. If the number of insects is high, then the results may not be anywhere near as good as they are when

### Bug juice

A number of gardeners using organic principles capture the pests that are causing problems and put them through the household blender. They then add water and spray this mixture over the plants. Apparently other bugs do not find treated plants as desirable as they were before treatment with the residues of their fellows. While I do know that this works, I'm not sure that I'd be too keen on having a milkshake mixed up in the blender after it has been used for this purpose. Bug juices are available commercially, but usually only in larger horticultural or agricultural packs.

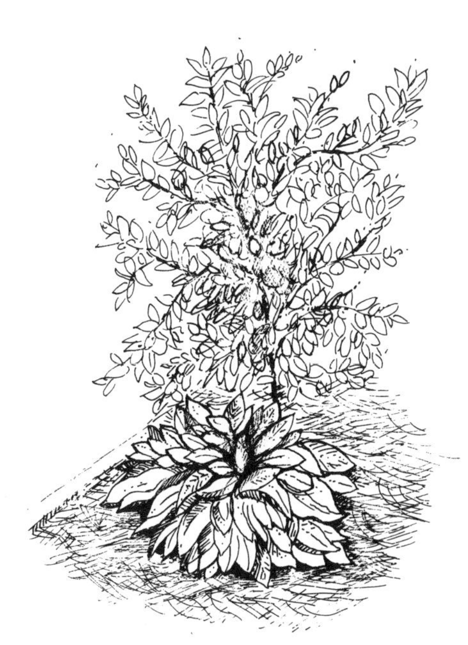

**comfrey planted under a fruit tree**

Companion planting relies on the repellent effects of herbs and other plants on certain insects; for example, nasturtiums are said to repel aphids; marigolds are used to discourage nematodes and comfrey is often used as a decoy to protect valued plants from grasshoppers.

Wormwood repels fruit fly and other pests but should not be planted close to the cultivated plants, since it is not a good companion. While it repels pests, it is also said to inhibit growth. Always plant wormwood in a border around the edges of the vegetable or fruit patch, rather than in among your valued plants.

there are only a few insects about. Similarly I do not believe that herbs have the same levels of fragrance in warmer climates, where the insect pressures are usually higher, as they do in colder climates.

Having said that though, I still consider it worthwhile to try companion planting as a means of reducing the problems of pests and diseases.

As already mentioned, the French and African Marigolds are well known as companion plants used to repel nematodes in tomatoes, capsicums, potatoes and other crops.

Orange-flowered nasturtiums are said to repel aphids. A gardener I know has these planted under all her roses and not an aphid is to be found on them. Nasturtiums grown near cabbage, cauliflower, kohlrabi and turnip are said to keep aphids away from these crops as well. Apparently the aphids are attracted to the nasturtiums rather than the vegetables, so they leave the vegetables alone. Comfrey acts in a similar manner as far as grasshoppers are concerned, and a bed of comfrey will help keep the vegetables and ornamentals free from attack. Always remember, though, to keep one bed set aside for comfrey rather than growing it in various parts of the garden, as it will spread quite readily.

Members of the *Artemisia* family, the wormwoods and southernwoods, are said to repel both cabbage moth and fruit fly, but because they are bad companions in terms of their growth habits, it is important to plant a barrier of these plants around the perimeter of the fruit and vegetable patch, rather than in among the valued plants.

Garlic and garlic chives are both said to repel aphids so these can be quite comfortably grown with plants that are likely to be affected by aphids. Many a good rose garden has garlic or garlic chives growing merrily throughout the bed.

Pennyroyal and tansy also repel insects; for example, tansy is said to repel cutworms, which chew through the stems of seedlings.

Gardeners have tried all these companion plantings over the centuries and, while I would not suggest that gardeners in the warmer parts of Australia should rely entirely on these control methods, I would suggest they give them a try. Be prepared, however, for some damage as this is only natural in an organically grown garden.

Comfrey is useful as a compost activator, but is also used to attract grasshoppers. These pests tend to feed on the foliage of the comfrey and leave the other vegetables alone. The comfrey appears to be able to cope quite well with the voracious feeding habits of grasshoppers.

## Non-chemical treatments

Never be fooled into believing that there is any such thing as a 'safe' insecticide or fungicide. Remember the importance of the predators, and remember also that the so-called 'safe' insecticides and fungicides invariably destroy predators at the same rate as the so-called 'unsafe' ones.

### Commercially available products

#### Pyrethrum

Pyrethrum is a natural extract of various members of the daisy family (*Chrysanthemum* and *Pyrethrum* genera). The extract is a broad-spectrum insecticide which is non-toxic to mammals but controls a wide range of pests. It has an efficacy of about twelve hours and does not persist on the plant after that time. This, of course, is one of its virtues, and contributes to its short withholding period.

Commercial crops of daisies are grown for pyrethrum extraction and these extracts are the ones most commonly available from garden shops. Pyrethrum on its own is not as effective as when it is combined with piperonyl butoxide, as it is in most home garden packs.

Scientists discovered that they were able to synthesise pyrethrum in the laboratory and these are now available on the market and are described as 'pyrethrins'. They have the same efficacy as natural pyrethrum.

A further step in the scientific chain is the 'pyrethroid', a man-made product derived from pyrethrins. Although based on organic principles, it is chemically synthesised. Pyrethrums and pyrethrins control aphids, most other sucking insects and the majority of chewing insects, but remember that they are absolute dynamite on bees and the predatory insects. The pyrethroids, on the other hand, are generally less harmful to bees.

*Chrysanthemum cinerariifolium* is a common source of pyrethrum, a broad-spectrum insecticide which controls a wide range of pests and is non-toxic to mammals.

## Derris dust

Derris dust has the active ingredient 'rotenone' and is a powder extracted from the roots of various South American plants. This, too, is a broad-spectrum insecticide that controls hard- and soft-bodied insects. It acts as a stomach poison and is effective for about 48 hours. Being a dust, it is difficult, to say the least, to apply to the underside of plants; however it is easy to apply to the upper surfaces of plants and vegetables such as cabbages and cauliflowers. It does a wonderful job of controlling the chewing and sucking insects that attack these plants.

Like pyrethrins, it is toxic to bees and therefore should not be applied to flowers. It will control the Corn Earworm when dusted onto the silk of the corn. It will wash off readily in rain and therefore needs to be reapplied. It has a withholding period of only one day, making it eminently suitable for the control of vegetable pests.

## *Bacillus thuringensis*

*Bacillus thuringensis* is a microbial biological control organism effective against the leaf-eating caterpillars. It acts as a stomach poison by forming a protein crystal within the stomach of the insect. It is specific to caterpillars only and will not harm any other warm-blooded insect, animal or fish. It is non-toxic to plants as well. It is normally sold as 'Dipel'.

Home-made garlic and onion spray will kill chewing and sucking insects on contact, but it may also harm bees and other predators.

## Garlic spray

Garlic spray has long been used as a control for a wide range of insects. It has a relatively quick knock-down action and controls both sucking and chewing insects. It is, however, dangerous to bees and other predators, and should be used with care.

## Eucalyptus oils

Eucalyptus oils are also used in the control of insects and the same comments apply as have been made for garlic spray.

## Potassium salts

The potassium salts are used as insecticides in the various soap sprays on the market. These are used to control aphids and mites, as well as a number of other sucking insects. These, too, have a toxic effect on bees.

## Fruit fly traps

A number of fruit fly traps are available and these come with various different attractants. In some cases the attractant is a pheromone, the fragrance emitted

**Garlic and onion spray**

Chop up 4 hot chillies, 4 large onions and 2 cloves of garlic and cover these with warm, soapy water. Remember to use Sunlight soap, not detergent. Leave to stand for 24 hours and then strain. Add a further 5 litres of water to the strained liquid. This will kill both chewing and sucking insects on contact, but it may also harm bees and other predators. If kept in a sealed container in a dark cupboard, the solution will keep for some weeks.

Always gather up dead or diseased foliage and fruit and place in a plastic bag. Before disposing of it, tie the top tightly and leave in the sun for as long as possible to destroy diseased organisms and insects. Never put affected material in the compost heap or bin.

by the female fruit fly when enthused about the mating process. These lure the male fruit flies into the trap, where they are often killed by an insecticide.

Other traps use baits of protein, often incorporated in an insecticide mixture, to attract both male and female fruit flies, while others are like the sticky fly paper of yesteryear, only in this instance the glue is inside a yellow, tent-like apparatus. All of these lures and baits are hung in the fruit trees.

### Seaweed extracts

Seaweed extracts are primarily sold as fertilisers but there are a number of organic gardeners who are convinced that regular spraying with these products will keep the fungal diseases, particularly the mildews, at bay, so this may be worth a try.

## Home-made recipes

There are many home recipes which can easily be made from common ingredients and most have a long history of effectiveness. Individual recipes for these are given beside the section on the particular pests and diseases which they control.

### Washing or baking soda

These are said to control mildews, rusts and scale. Mix 100 grams of soda in 4.5 litres of water, add some Sunlight soap and spray over susceptible plants.

### White Oil

To make White Oil, place a cup of domestic cooking oil in a blender, add one and a half cups of water and a teaspoon of Sunlight soap and blend. Store in a tightly sealed jar. Dilute at the rate of one part of oil to ten parts of water.

### Insect-repellent herbs

| Herbs | Insects repelled |
| --- | --- |
| garlic, garlic chives | aphids |
| marigolds | nematodes |
| pennyroyal | ants |
| rue | snails, slugs |
| sage | Cabbage Moth |
| tansy | ants, cutworm |
| wormwoods | Cabbage Moth, Fruit fly, snails and slugs |

## Hygiene

The alternatives to chemical spraying for pest and disease control are virtually limitless, but the most important of all is impeccable garden hygiene.

Always gather up dead or diseased foliage or fruit and place in a plastic bag, tie the top tightly and leave in the sun for as long as possible in order to destroy the diseased organisms. Insects will be killed in the same manner. Because it is unlikely that compost heaps will generate sufficient heat to kill the troublesome pests and diseases, it is best not to place affected material in the compost heap or bin.

Weed control is also important because, as we have already seen, many of our harmful pests and diseases overwinter in weeds around fence lines and in waste areas.

### Hand removal of insects

Hand removal of many insects can be a most effective method of pest control, but it is a good idea to wear gloves when carrying out this task. It is easier to cut grasshoppers in half with scissors or secateurs if you are wearing gloves, simply because they prevent the grasshopper from smelling the presence of a human hand.

## Conclusion

It is the gardener's responsibility to learn more about the pests and diseases that attack our gardens and, therefore, to adopt a more responsible attitude towards the use of pesticides and fungicides. Identification is the first step in that responsibility and will enable you to take the appropriate action. Sometimes the most appropriate action may simply be to do nothing at all. I sincerely hope that, if enough gardeners adopt the practices recommended in this book, our gardens will be a more harmonious place for all the creatures of nature.

# Table of pesticides and fungicides

(I) = insecticide; (F) = fungicide

| Active Ingredient | Trade name | Features |
|---|---|---|
| *Bacillus thuringensis* (I) | Caterpillar Killer<br>Dipel | Biological caterpillar killer |
| Carbaryl (I) | Carbaryl insecticide (produced by various manufacturers) | Non-systemic control of chewing insects<br>Controls sucking insects on contact |
| Chlorothalonil (F) | Agchem Garden Fungicide<br>Garden King Lawn Garden Fungicide | Non-systemic fungicide for the control of a wide range of fungal problems in gardens and lawns |
| Chlorpyrifos (I) | Garden King Antkil<br>Chemspray Chlorban<br>Agchem Fix-Ant<br>Garden King Grubkil | Non-systemic control of ants, cockroaches and a wide range of other pests of house, garden and lawn |
| Copper (F) | Hortico Bordeaux<br>Agchem Bordeaux<br>Multicrop Kocide<br>Garden King Copper Spray<br>Defender Curly Leaf Fungicide<br>Hortico Leaf Curl Spray | A non-systemic fungicide widely used for centuries for most common fungi |
| Cyfluthrin (I) | Baythroid Lawn Grub Insecticide | A synthetic pyrethroid for the control of lawn pests |
| Diazinon (I) | Hortico Ant Killer<br>Garden King Diazamin<br>Hortico Lawn Grub & Insect Killer | A penetrative insecticide for the control of ants, leaf miners and many chewing insects |
| Dicofol (I) | Hortico Kelthane | A miticide |
| Dimethoate (I) | Chemspray Rogor<br>Hortico Rogor<br>Garden King Rogor<br>Defender Rogor | A systemic insecticide widely used to control sucking insects |
| Fenamiphos (I) | Bayer Lawn Beetle Killer<br>Bayer Nemacur | For the control of soil and lawn pests |
| Fenthion (I) | Lebaycid | A penetrative insecticide for the control of a wide range of insects, including fruit fly |
| Fluvalinate (I) | Yates Mavrik | A synthetic pyrethroid which controls a wide range of pests<br>This product is not toxic to bees |
| Maldison (I) | Hortico Malathion<br>Garden King Malathion | Non-systemic, controls a wide range of chewing and sucking insects |

(I) = insecticide; (F) = fungicide

| Active Ingredient | Trade name | Features |
|---|---|---|
| Mancozeb (F) | Yates Mancozeb Plus<br>Chemspray Mancozeb | Non-systemic, controls a wide range of fungal diseases |
| Omethoate (I) | Folimat | Systemic insecticide for the control of a wide range of insects, particularly sap suckers |
| Petroleum Oil (I) | Hortico Clear White Oil<br>Garden King White Oil<br>Defender All Season Oil Insecticide<br>Chemspray Clear White Oil | A suffocant used mostly for the control of scale insects |
| Pyrethrum (I) (Pyrethrins) | Various manufacturers use pyrethrum or pyrethrin, usually in combination with piperonyl butoxide | A non-systemic organic insecticide for the control of chewing and sucking insects |
| Rotenone (I) | Derris Dust | A non-systemic organic insecticide for the control of a wide range of pests |
| Sulphur (F) | Garden King Sulphur Spray<br>Various manufacturers incorporate sulphur with other ingredients to form multi-purpose products | An inorganic compound for the control of fungi.<br>It also has action against some mites |
| Triadimefon (F) | Bayleton Garden Fungicide | A fungicide, particularly effective against petal blight |
| Trichlorfon (I) | Bayer Caterpillar Killer<br>Bayer Lawn Grub Killer<br>Defender Lawn Grub Killer | For the control of caterpillars and soil pests |
| Triforine (F) | Kendon Triforine (fungicide only)<br>Hortico Black Spot & Insect Killer<br>Zest Rose Spray<br>Defender Rose Black Spot & Insect Spray | Systemic fungicide for the control of black spot and powdery mildew. Widely used in mixed sprays for roses |
| Zineb (F) | Chemspray Zineb | A broad spectrum fungicide |

## For New Zealand gardeners

The following products are commonly available to New Zealand gardeners:

(I) = insecticide; (F) = fungicide

| Active Ingredient | Trade name | Features |
|---|---|---|
| Acephate (I) | Yates Orthene | A systemic insecticide |
| Clorothalonil (F) | Coopers General Fungicide | A general-purpose fungicide |
| Chlorpyrifos (I) | Yates Soil Insect Killer | A granular insecticide for the control of soil-inhabiting insects |
| Diazinon (I) | Yates Target | A broad-spectrum insecticide |
| Dicofol (I) | Yates Mite Spray | A broad-spectrum miticide |

These are only a few of the more common insecticides and fungicides on the market today. Remember that all home garden chemicals must be registered and you are therefore only allowed to use them for the purposes for which they are registered. This list should be taken as a guide only. Study the label of particular products to find out whether they are registered for use in your state.

# Further Reading


Conacher, J., *Pests, Predators and Insecticides*, Organic Growers Association, West Australia, 2nd Edition, Wembley, 1980

Cundall, P., *Organic Gardening*, Gardening Australia Collectors series No 1. Federal Publishing, Sydney, 1992

French, J., *The Organic Garden Doctor*, Angus and Robertson, Sydney, 1988

Gross, J., *The Garden Doctor*, Kangaroo Press, Kenthurst 1983, 1992

Hamilton, G., *Successful Organic Gardening*, Macmillan, Melbourne, 1987

Jones, D. and Elliot, R., *Pests, Diseases and Ailments of Australian Plants*, Lothian, Melbourne, 1986

McMaugh, J., *What Garden Pest or Disease Is That?*, Lansdowne, Melbourne 1985

Smith, K., *The Backyard Organic Garden*, Lothian, Melbourne, 1990

Swaine, Ironside and Corcoran (Eds), *Insect Pests of Fruit and Vegetables*, Queensland Dept of Primary Industries, Brisbane, 1991

Waters, D., *Simple Insect Control for the Home Garden Using Herbs*, Pennyroyal Herb Farm, Bundaberg, 1986

Yates, Arthur Ltd, *Yates Garden Guide*, Collins, Sydney, 1992

Yates New Zealand Ltd, *Yates Garden Doctor*, Auckland, 1990

# Index


African black beetle 10, 17
anthracnose 43
ants 17, 32
aphids 7, 18
apple scab 43
Azalea Lace Bug 18
azalea leaf miner 18

bacteria 41
bacterial insecticides 16
bean fly 11, 19
black spot 44
blossom end rot 44
borax 15
Bordeaux mixture 35
borers 10, 19
broad mite 19, 29
Bronze Orange Bug 20
brown patch 47
brown rot 45
bud worms 20
bug juice 56
bulb, stem and root attackers 10

cabbage white butterfly 20
caterpillars 8, 21
chamomile tea 45
Cineraria Leaf Miner 8
Citrus Gall Wasp 8, 11, 21
Citrus Leaf Miner 8, 11
citrus scab 45
chewing insects 8, 18, 20, 21, 25
Christmas Beetle 21
club root 37
Condy's crystals 45
Codling Moth 11, 22
collar rot 45
companion planting 57
Corbie 22
Corn Earworm 22
crop rotation 56
cutworms 23
Cyclamen Mite 29

disease prevention 49, 60
Derris Dust 16, 58
dollar spot 47
downy mildew 46

earwigs 23
Erinose Mite 23
eucalyptus oil spray 59
*Eucalyptus torelleana* 31
*Eoudia elleryana* 23

Figleaf Beetle 23
figs 23
fruit fly 5, 11, 24, 59
fruit fly baits 15
fungi 5, 41

gall-forming insects 8, 11
garlic spray 59; and onion spray 59
gladioli thrips 24
grape phylloxera 25
grasshoppers 25
grease bands 13
Great Potato Blight 35
Green Vegetable Bug 26

Harlequin Bug 26
Hawk Moth Caterpillar 26
hibiscus 23
Hibiscus Beetle 27
honeydew 17, 31, 32
hygiene 60

Ichneumon Wasp 28
insect-attracting plants 33
insect-repellent herbs 60

ladybirds 28
lantana leaves 13
lawn diseases 47; grubs 27
leaf blight 45
Leaf Blister Sawfly 28
leaf curl 46
leaf spot 46
lichen 41
longicorn beetle 8
lychees 23

Macadamia Nut Borer 29
Maori Mite 29
mealy bugs 7, 29
methylated spirits 14
mildews 46
mites 19, 29
molasses 16
Monolepta Beetle 31
mosaic virus 47
mycorrhiza 5

nematodes 39
nut borer 29
nutrient deficiency symptoms 36

Orange Palm Dart 30

Palm Dart 30
parasitic plants 40
Pear and Cherry Slug 30
petal blight 47
pyrethrins 16
pyrethrum 16, 59
physiogenic diseases 36
phytophthora 48
phytotoxicity 38
poison schedule 50
powdery mildew 46
psyllids 31
pythium 48
Pumpkin Beetle 31

Red-shouldered Leaf Beetle 31
Red Spider Mite 29
resistance factor 55
rhubarb spray 14
rust 47

safety 51, 52
salts, excess 37
Sawfly 28
seaweed extracts 60
scale insects 32
snails and slugs 32
soda, washing and baking 60
soil diseases 48
sooty mould 10
spiders 25
stomach poisons 12
storage, chemical 53
sucking insects 9, 10, 19, 20
synthetic pyrethroids 16
systemic poisons 13
Syzygiums 31

thrips 33
tissue feeders 11, 18
tomato wilt 48
Two-Spotted Mite 29

virus 30, 38, 47; transmission 38

webbing caterpillars 33
white oil, home-made 60
withholding period 52